PRAISE FOR *FACTOR FOUR*

'*Factor Four* made my spine tingle. This is a book which captures the moment, which crystallises a set of seemingly conflicted trends and issues around a simple idea... carries you away with its optimism and can-do spirit, its hard facts and concrete examples' *TOMORROW*

The message of a radical, profitable and sustainable change comes through loud and clear... perceptive and exciting ideas for transforming our entire economic system' *NATURE*

'By any criteria a key "text for the millenium"... for anyone involved in business and the environment' *GREENER MANAGEMENT INTERNATIONAL*

'It would be a triumph of humanity over itself if the world were to embrace the practical model of sustainability set out in this book' *RSA JOURNAL*

'This excellent book... [is for] anyone with a practical bent and a real interest in changing the world for the better' *RESURGENCE*

'Probably the cleverest book on the business–environment–society relationship... superb, and deserves every ounce of the accolades it has received' *SOCIAL AND ENVIRONMENTAL ACCOUNTING*

'An exciting challenge' *THE ENGINEER*

'An engaging, challenging work, highly-recommended for the interested, non-specialist reader' *THE BOOKWATCH*

'A startlingly simple message that offers a new way of looking at progress, moving the emphasis from labour productivity to resource productivity' *ECODESIGN*

'Well-documented, punchy and brilliantly incisive, full of information and original proposals' *PEOPLE AND THE PLANET*

'This book is so entertaining that it can absorb the reader in the most distracting situations... it wonderfully succeeds in showing what is possible, how important it is to change, and how much fun is to be had from all this. It is the best advertisement for the future I have ever come across' *BUILDING DESIGN*

'Seldom can the venerable Club of Rome have received so passionate a report, or one so dismissive of the prevailing direction of world leaders and businessmen' *THE GUARDIAN*

Ernst von Weizsäcker is President of the Wuppertal Institute for Climate, Environment and Energy in the North Rhine/Westphalian Science Centre, Germany. He was previously a Professor of Biology at Essen University, President of Kassel University, Director at the UN Centre for Science and Technology for Development (New York) and Director of the Institute for European Environmental Politics in Bonn, and in 1996 he was the first recipient of the Duke of Edinburgh Gold Medal of WWF International.

Amory B Lovins and L Hunter Lovins are respectively Vice President and Director of Research, and President and Executive Director, of Rocky Mountain Institute which they co-founded in 1982 in Colorado, USA. Amory Lovins is a MacArthur Fellow and the recipient of six honorary doctorates, having been perhaps the youngest-ever Oxford don. Hunter Lovins is a sociologist, political scientist and barrister, and was Henry R Luce Professor of Environmental Studies at Dartmouth College. The Lovinses have been joint recipients of the Mitchell Prize, the Right Livelihood Award (the 'alternative Nobel Prize'), the Onassis Foundation's first Delphi Prize and the Nissan Prize. Their pioneering technical, economic and policy work has been enormously influential in the energy, construction and car manufacturing industries.

Factor Four

Doubling Wealth – Halving Resource Use

THE NEW REPORT TO THE CLUB OF ROME

Ernst von Weizsäcker,
Amory B Lovins
and
L Hunter Lovins

For David Kupfer — just go do it!
Amory
7.XI.2000

EARTHSCAN PUBLICATIONS LTD, LONDON

First published in the UK in 1998 by
Earthscan Publications Limited

A catalogue record for this book is available from the British Library

ISBN: 1 85383 406 8

Page design and typesetting by PCS Mapping & DTP, Newcastle upon Tyne

Printed and bound by Clays Ltd, St Ives plc

Cover design by Andrew Corbett

For a full list of publications please contact:
Earthscan Publications Limited
120 Pentonville Road
London N1 9JN
Tel: (0171) 278 0433
Fax: (0171) 278 1142
Email: earthinfo@earthscan.co.uk
WWW: http://www.earthscan.co.uk

Earthscan is an editorially independent subsidiary of Kogan Page Limited and publishes in association with WWF-UK and the International Institute for Environment and Development.

Contents

Part Two: Making it Happen – Improving Profitability

Chapter 4: If Markets Create the Problem, Can They Also Provide Much of the Answer?

Chapter 5: Buying and Selling Efficiency

Chapter 6: Reward What We Want, Not the Opposite

Chapter 7: Ecological Tax Reform

Part Three: A Sense of Urgency

Chapter 8: The Challenge from Rio

Chapter 9: Avalanches of Matter: The Forgotten Agenda

List of Abbreviations

ACT^2	Advanced Customer Technology Test for Maximum Energy Efficiency
ACTS	Advanced Communications Technologies and Services
AOSIS	Alliance of Small Island States
Btu	British thermal units
ca	circa
CAFE	Corporate Average Fuel Economy
CFCs	chlorofluorocarbon(s)
cm	centimetre
CoP	Conference of Contracting Parties
DSD	Duales System Deutschland
ETR	ecological tax reform
FCCC	Framework Convention on Climate Change
GATT	General Agreement on Tariffs and Trade
GDP	gross domestic product
GNP	Gross National Product
GPI	Genuine Progress Indicator
hp	horsepower
ICC	International Chamber of Commerce
ICPD	International Conference on Population and Development
ICSU	International Council of Scientific Unions
IIASA	International Institute for Applied Systems Analysis
IMF	International Monetary Fund
INC	Intergovernmental Negotiating Committee
IPCC	Intergovernmental Panel on Climate Change
ISEW	Index of Sustainable Economic Welfare
ISTEA	Intermodal Surface Transportation Efficiency Act
MIPS	material inputs per service
NAFTA	North American Free Trade Agreement
NEF	New Economics Foundation

NSSL	National Seed Storage Laboratory
OECD	Organisation for Economic Cooperation and Development
OPEC	Organisation of Petroleum Exporting Countries
PCSD	President's Council for Sustainable Development
PG&E	Pacific Gas and Electric Company
PUCs	public utility commissions
PVC	polyvinyl chloride
ROI	return on investment
TPA	Third Party Access
UNCED	United Nations Conference on Environment and Development
UNEP	United Nations Environment Programme
vs	versus
WMO	World Meteorological Organisation
WRAP	Waste Reduction Always Pays
WTO	World Trade Organisation

List of Figures

Foreword

Factor Four is the right idea, at the right time, to become a symbol of progress, an outcome the Club of Rome would welcome. Doubling wealth while halving resource use is at the heart of what *The First Global Revolution* (King and Schneider, 1991) is demanding, the first-ever Report by the Club of Rome. Unless we manage to double wealth, how can we ever hope to solve the problems of poverty that Bertrand Schneider (1994) so compellingly addresses in *The Scandal and the Shame?* And how can we cope with the governability challenge elucidated by Yehezkel Dror in his recent report on governability?

On the other hand, how can we ever return to an ecological balance on earth unless we are able to cut resource use in half? Halving resource use truly means *Taking Nature into Account*, the title of Wouter van Dieren's most recent report to the Club. Halving resource use closely relates to the challenge of sustainable development which dominated the 1992 Earth Summit in Rio de Janeiro. But remember that this challenge was addressed 20 years earlier in the famous *Limits to Growth* Report to the Club of Rome by Dana and Dennis Meadows, Jørgen Randers and Bill Behrens (Meadows et al, 1972).

Doubling wealth and halving resource use thus jointly indicate the magnitude of the *world problematique* which the Club of Rome considers to be at the core of its activities. We are proud to be able to introduce *Factor Four* as an encouraging new Report to the Club that outlines some of the steps the world will have to take. *Factor Four* may contribute to the *resolutique* that the Club heralds in *The First Global Revolution.* We gratefully acknowledge the contribution of two of the world's leading pioneers in energy efficiency, Amory and Hunter Lovins, who were roped into the effort by our member Ernst von Weizsäcker, who initiated the process of making *Factor Four* another Report to the Club. It was also his idea to include the essential considerations of material resources and transportation. The team of authors managed to collect an impressive 50 examples of quadrupling resource productivity, thus demonstrating the broad feasibility of Factor Four.

Each Report to the Club of Rome is the end-point of an exhaustive process of research and discussion among club members and other leading experts. In the case of *Factor Four*, the process culminated in an international conference of the Club of Rome cosponsored by the Friedrich Ebert Foundation. Held in Bonn in March 1995, the meeting offered all interested club members the opportunity to provide input to the emerging report, of which a draft was made available in advance. The Executive Committee of the Club of Rome arrived at a positive decision in June 1995 to accept the amended manuscript as a Report to the Club.

In the name of the Club of Rome, I express my sincere hope that this new report will make its mark on the international discussion among policymakers and experts alike.

Ricardo Diez Hochleitner
President of the Club of Rome
Madrid, December 1996

Preface

This is an ambitious book which seeks to redirect technological progress. Aggressive increases in labour productivity constitute a rather questionable programme at a time when more than 800 million people are out of work. At the same time, scarce natural resources are squandered. If resource productivity were increased by a factor of four, the world could enjoy twice the wealth that is currently available, whilst simultaneously halving the stress placed on our natural environment. We believe that we can demonstrate that this quadrupling of resource productivity is technically feasible and would produce massive macro-economic gains, ie make individuals, firms and all of society better off.

As we outline in this groundbreaking programme, we are taking up the concerns expressed in the early 1970s by the Club of Rome, which shook the world with its report *The Limits to Growth* (Meadows et al, 1972). This time, however, we give an optimistic answer to these concerns. We will demonstrate that equilibrium scenarios are available. *Factor Four*, we believe, can put the Earth back into balance (to adapt a metaphor from Al Gore's challenging bestseller (Gore, 1992)).

We wish to thank the Club of Rome for a steady and growing interest in our project. A Club of Rome ad hoc seminar was organized in March 1995 in Bonn, co-sponsored by the Friedrich Ebert Foundation and the German Environment Foundation, to discuss the manuscript of the book. As a result, much of the text was rewritten and sent to members of the Club's Executive Committee, which in June 1995 accepted the book as a Report to the Club. We are honoured that the Club's President provided a Foreword to this edition.

The manuscript was originally written in various versions of

English – half of it by a native German speaker, half by two Americans who had lived for 2 and 14 years in England but had scarcely succeeded in learning English. For its first publication, the whole book was translated into German and presented in September 1995 as *Faktor Vier: Doppelter Wohlstand – halbierter Naturverbrauch* by Droemer-Knaur, Munich. (The subtitle translates loosely as 'Live Twice as Well, Use Half as Much' and more literally as shown on this book's title page.) It became a bestseller almost instantly, and remained so for more than six months. A Czech translation was published in 1996, and a Spanish translation in 1997. French, Italian, Japanese, Russian and Swedish editions are underway, and enquiries have been made about other languages. Industrial interest rapidly grew worldwide. The Carinthia Trade Fair in Klagenfurt, Austria, is planning an international Factor Four Fair from 16–18 June 1998. And hundreds of letters of encouragement have been written to the authors, many of which offered further practical examples of the principles of *Factor Four*. Moreover, two of us – Amory B Lovins and L Hunter Lovins – are preparing a complimentary book, with Paul Hawken, framed for the US rather than the European context and primarily for a business audience.[1]

We are greatly indebted to all those who were engaged in the discussion of this book even before it became available in what we hope is a more faithful version of the English language. Hundreds of people have been involved in the making of this book. Here we name but a few of them, including those who actively participated in the Club of Rome's review meeting: Franz Alt, Owen Bailey, Benjamin Bassin, Maryse Biermann, Jérôme Bindé, Raimund Bleischwitz, Stefanie Böge, Holger Börner, Hartmut Bossel, Frank Bosshardt, Stefan Bringezu, Leonor Briones (Manila), Bill Browning, Michael Brylawski, Maria Buitenkamp, Scott Chaplin, David Cramer, Maureen Cureton, Hans Diefenbacher, Wouter van Dieren, Ricardo Díez Hochleitner, Reuben Deumling, Hans Peter Dürr, Barbara Eggers, Felix FitzRoy, Claude Fussler, Paul Hawken, Rick Heede, Peter Hennicke, Friedrich Hinterberger, Alice Hubbard, Wolfram Huncke, Reimut Jochimsen, Ashok Khosla, Albrecht Koschützke, Sascha Kranendonk, Hans Kretschmer, Martin Lees, André Lehmann, Harry Lehmann, Christa Liedke, Jochen Luhmann, Manfred Max-Neef (Valdivia), Mark Merritt, Niels Meyer, Timothy Moore, Kikujiro Namba (Tokyo), Hermann

1 Paul Hawken, Amory B Lovins and L Hunter Lovins (to be published in 1998) Natural Capitalism Earthscan Publications Ltd, London

Ott, Andreas Pastowski, Rudolf Petersen, Richard Pinkham, Wendy Pratt, Joseph Romm, Jen Seal, Wolfgang Sachs, Karl-Otto Schallaböck, Friedrich Schmidt-Bleek, Harald Schumann, Eberhard Seifert, Farley Sheldon, Bill Shireman, Walter Stahel, Klaus Steilmann, Ursula Tischner, Reinhard Ueberhorst, Carl Christian von Weizsäcker, Christine von Weizsäcker, Franz von Weizsäcker, Anders Wijkman and Heinrich Wohlmeyer.

Without the pioneering work of Herman Daly, Dana and Dennis Meadows, Paul Hawken, Hazel Henderson, Bill McDonough and David Orr, it would have been almost impossible to conceive of a book with this scope.

We also thank the sponsors of the Bonn meeting and the government of North Rhine Westphalia for the basic grant given to the Wuppertal Institute for Climate, Environment and Energy, as part of the North Rhine Westphalian Science Centre, with the injunction to investigate and put into practice the principles of this book.

Not least, our thanks go to Earthscan Publications in London who have edited and promoted the book excellently. Special thanks go to Jonathan Sinclair Wilson and Rowan Davies.

Ernst von Weizsäcker
Amory B Lovins
L Hunter Lovins
January 1997

Introduction
More for Less

Exciting Prospects for Progress

'Factor Four', in a nutshell, means that resource productivity can – and should – grow fourfold. The amount of wealth extracted from one unit of natural resources can quadruple. Thus we can live twice as well – yet use half as much.

That message is both novel and simple.

It is novel because it heralds nothing less than a new direction for technological 'progress'. In the past, progress was the increase of labour productivity. We feel that *resource productivity* is equally important and should now be pursued as the highest priority.

Our message is simple, offering a rough quantitative formula. This book describes technologies representing a quadrupling or more of resource productivity. Progress, as we have known at least since the Earth Summit at Rio de Janeiro, must meet the criterion of sustainability. *Factor Four* progress does.

The message is also *exciting*. It says that some aspects of that efficiency revolution are available now at *negative* cost; that is, profitably. Much more can be made profitable. Countries engaging in the efficiency revolution become stronger, not weaker, in terms of international competitiveness.

That is not only true for the industrialised countries of the North. It is even more valid for China, India, Mexico or Egypt – countries that have a great supply of inexpensive labour but are short of energy. Why should they learn from the US and from Europe how to waste energy and materials? Their journey to prosperity will be smoother, swifter and safer if they make the efficiency revolution the centrepiece of their technological progress.

The efficiency revolution is bound to become a global trend. As is always the case with new opportunities, those who pioneer the trend will reap the greatest rewards.

Moral and Material Reasons

Changing the direction of progress is not something a *book* can do. It has to be done by people – consumers and voters, managers and engineers, politicians and communicators. People don't change their habits unless they have good reasons for doing so. Motivation needs to be experienced as compelling and urgent by a critical mass of people, otherwise there won't be enough momentum to change the course of our civilisation.

The reasons for changing the direction of technological progress are both moral and material. We trust that most readers share our view that preserving physical support systems for humankind is a high moral priority. The ecological state of the world demands swift action as will be discussed in Part III. We avoid the language of doom and gloom, but we do present some disturbing ecological facts and trends. These should be established if we want to say something in quantitative terms about the necessary answers. We shall demonstrate that gaps as large as a factor of four are opening before us which need to be closed (see Figure 1, pxxviii).

If these gaps aren't closed, the world may run into unprecedented troubles and disasters. Avoiding them may seem like a formidable task. Can such gigantic gaps be closed at all? This question leads us to the good news. The gaps can be closed. *Factor Four* is at the heart of the answer. Best of all, we shall discover very strong *material* reasons for changing the direction of technological progress.

Countries starting at once will reap major benefits. Countries that hesitate are likely to suffer formidable losses of their capital stock which will quickly become obsolete as resource efficiency trends take hold elsewhere.

Efficiency Cure for the Wasting Disease

Why do we believe this? Essentially because we see our society in the grip of a severe but curable illness. It is not unlike the disease our grandparents called 'consumption' because it made its victims waste away. Today's economic tuberculosis consumes neither our bodies nor our resources (used energy and resources stay behind as unproductive pollution), but its effect on people, nations and the planet is just as deadly, costly and contagious.

We have been told that industrialisation has resulted from increasing levels of efficiency and productivity. Human productivity has certainly multiplied manyfold since the beginning of the Industrial Revolution. We have increased our productive capacities by substituting resources for human labour. Yet that substitution has now gone too far, over-using such resources as energy, materials, water, soil and air. Gains in 'productivity' pursued in this way are thus overwhelming the living systems that provide our resource base and must also assimilate the detritus of our civilisation.

A currently popular line of rhetoric maintains that any solution to these environmental problems will be very costly. That is wrong. What makes it wrong is the revolution in resource efficiency that this book is all about. Correcting the imbalances between how we use people and resources, improving resource efficiency and healing the 'wasting disease' represent, in fact, major economic opportunities. Much of the cure is not painful but soothing to both natural systems and the social fabric of global civilisation.

When people think of waste, they consider their household garbage, exhaust gases from their cars and the containers of rubbish outside businesses or construction sites. If you were to ask how much material is wasted each year, most people would admit that a certain percentage is wasted, but not a great deal. Actually, we are more than ten times better at wasting resources than at using them. A study for the US National Academy of Engineering found that about 93 per cent of the materials we buy and 'consume' never end up in saleable products at all. Moreover, 80 per cent of products are discarded after a single use, and many of the rest are not as durable as they should be. Business reformer Paul Hawken estimates that 99 per cent of the original materials used in the production of, or contained within, the goods made in the US become waste within 6 weeks of sale.

Most of the energy, water and transportation services we consume are wasted too, often before we get them; we pay for them, yet they provide no useful service. The heat that leaks through the attics of poorly insulated homes; the energy from a nuclear or coal-fired power station, only 3 per cent of which is converted into light in an incandescent lamp (70 per cent of the original fuel energy is wasted before it gets to the lamp, which in turn converts only 10 per cent of the electricity into light); the 80–85 per cent of a car's petrol that is wasted in the engine and drivetrain before it gets to the wheels; the water that evaporates or dribbles away before it gets to the roots of a crop; the senseless movement of goods over huge distances for a result equally well achieved more locally – these are all costs without benefits.

This waste is unnecessarily expensive. The average American, for example, pays nearly US$2,000* a year for energy, either directly purchased for the household or embodied in businesses' goods and services. Add to that wasted metal, soil, water, wood, fibre and the cost of moving all these materials around, and the average American is wasting thousands of dollars every year. That waste, multiplied by 250 million people, yields at least a trillion dollars per year that is needlessly spent. Worldwide, it may even approach US$10 trillion, every year. Such waste impoverishes families (especially those with lower incomes), reduces competitiveness, imperils our resource base, poisons water, air, soil and people, and suppresses employment and economic vitality.

The Efficiency Cure

Yet the wasting disease is curable. The cure comes from the laboratories, workbenches and production lines of skilled scientists and technologists, from the policies and designs of city planners and architects, from the ingenuity of engineers, chemists and farmers, and from the intelligence of every person. It is based on sound science, good economics and common sense. The cure is using resources efficiently; doing more with less. It is not a question of going backward or 'returning' to prior means. It is the beginning of a new industrial revolution in which we shall achieve dramatic increases in resource productivity.

Ways of doing this have significantly increased in the past few years, opening up wholly unexpected opportunities for business and society. This book is an introduction, description and call to action on behalf of these opportunities in advanced resource efficiency. It shows practical, often profitable ways to use resources *at least four times as efficiently as we do now*. Or to put it another way, it means we can accomplish everything we do today as well as now, or better, with only one-quarter of the energy and materials we presently use.

* Currency is given in US dollars unless otherwise indicated.

This would make it possible, for example, to double the global standard of living while cutting resource use in half. Further improvements on an even more ambitious scale are also rapidly becoming feasible and cost-effective.

Doing more with less is not the same as doing less, doing worse or doing without. Efficiency does not mean curtailment, discomfort or privation. When several presidents of the US proclaimed that 'energy conservation means being hotter in the summer and colder in the winter', they were not talking about energy *efficiency*, which should make us *more* comfortable by improving buildings so that they provide better comfort whilst using less energy and less money. To avoid this common confusion, this book avoids the ambiguous term 'resource conservation' and instead uses 'resource efficiency' or 'resource productivity'.

Seven Good Reasons for Resource Efficiency

We have given somewhat abstract moral and material reasons for moving into efficiency. Now we will become more concrete by offering seven compelling reasons for doing exactly that.

1) *Live better.* Resource efficiency improves the quality of life. We can see better with efficient lighting systems, keep food fresher in efficient refrigerators, produce better goods in efficient factories, travel more safely and comfortably in efficient vehicles, feel better in efficient buildings, and be better nourished by efficiently grown crops.
2) *Pollute and deplete less.* Everything must go somewhere. Wasted resources pollute the air, water or land. Efficiency combats waste and thus reduces pollution, which is simply a resource out of place. Resource efficiency can greatly contribute to solving such huge problems as acid rain and climatic change, deforestation, loss of soil fertility and congested streets. By themselves, energy efficiency plus productive, sustainable farming and forestry practices could make up to 90 per cent of today's environmental problems virtually disappear, not at a cost but – given favourable circumstances – at a profit. Efficiency can buy much time in which we can learn to deal thoughtfully, sensibly and sequentially with the world's problems.

3) *Make money.* Resource efficiency is usually profitable: you don't have to pay now for the resources that aren't being turned into pollutants, and you don't have to pay later to clean them up.
4) *Harness markets and enlist business.* Since resource efficiency has the potential of being profitable, much of it can be implemented largely in the marketplace, driven by individual choice and business competition, rather than requiring governments to tell everyone how to live. Market forces can theoretically drive resource efficiency. However, we are still confronted with the considerable task of clearing barriers and reversing daft incentive structures that keep the market from working fully.
5) *Multiply use of scarce capital.* The money freed up by preventing waste can be used to solve other problems. Developing countries in particular, with less of their capital sunk in inefficient infrastructure, are in an excellent position to multiply the use of scarce capital. If a country buys equipment to make very energy-efficient lamps or windows, it can provide energy services with less than a tenth of the investment that would be required to buy more power stations instead. By also recovering that investment at least three times faster and reinvesting it elsewhere, the services rendered by the invested capital can rise more than 30-fold. (Some calculations show savings far higher still.) For a good many developing countries this could be the only realistic way to achieve prosperity in a reasonable timespan.
6) *Increase security.* Competition for resources causes or worsens international conflict. Efficiency stretches resources to meet more needs, and reduces unhealthy resource dependencies that fuel political instability. Efficiency can reduce international sources of conflict over oil, cobalt, forests, water – whatever someone has that someone else wants. (Some countries pay in military costs, as well as directly, for their resource dependence: one-sixth to one-quarter of the US's military budget is earmarked for forces whose main mission is getting or keeping access to foreign resources.) Energy efficiency can even indirectly help block the spread of nuclear bombs by providing cheaper and inherently non-military substitutes for nuclear power plants and their dual-purpose materials, skills and technologies.
7) *Be equitable and have more employment.* Wasting resources is the other face of a distorted economy that increasingly splits society into those who have work and those who don't. Either way, human energy and talent are being tragically misspent.

> Yet a major cause of this waste of people is the wrong and profligate thrust of technological progress. We are making ever fewer people more 'productive', using up more resources and effectively marginalising one-third of the world's workforce. We need a rational economic incentive that allows us to employ more people and fewer resources, solving two critical problems at the same time. Businesses *should sack the unproductive kilowatt-hours, tonnes and litres rather than their workforce*. This would happen much faster if we taxed labour less and resource use correspondingly more.

This book presents a toolbox of modern resource efficiency. Fifty examples are presented of at least quadrupling resource productivity. Through these examples you can see for yourself what kinds of tools are available, how they work, what they can do, and how to apply them practically and profitably. Each of us – whether in business, the home or education, in the private, public or nonprofit sector, in concert with others or in our individual lives – can pick up the tools and get to work.

WHAT'S SO NEW ABOUT EFFICIENCY?

Efficiency is a concept as old as the human species. Much human progress, in all societies, has been defined by new ways to do more with less. A great deal of engineering and business is about using all kinds of resources more productively, but for the past 150 years much of the technological effort of the industrial revolution has been devoted to increasing *labour productivity, even if that required more generous use of natural resources*. Recently, however, resource efficiency has undergone such a conceptual and practical revolution that most people haven't yet heard about its new potential.

Ten years ago, the main news about how to do more with less was the speed of technological improvement. Since the oil crisis of the 1970s, we have learnt in each half-decade how, in principle, we could use electricity about twice as efficiently as before. Each time, that doubled efficiency would theoretically cost two-thirds less. Similar progress continues today – less through new technologies than through better understanding of how to choose and combine existing ones. Progress in making resource efficiency bigger and cheaper is thus quite dramatic. It is, in fact, somewhat comparable to the computer and consumer electronics revolution, where everything is continually becoming smaller, faster, better and cheaper. But the energy and material resource experts typically have not yet begun to think in terms of ever-increasing energy efficiency. The talk in official energy policy organisations seems still to concentrate on questions such as how much coal should be displaced by nuclear power and at what price – questions of supply that the demand-side revolution increasingly renders outdated and irrelevant.

The prejudice remains widespread that saving more energy will always cost more. The usual belief is that, beyond the familiar zone of 'diminishing returns', there will be a wall beyond which further savings are prohibitively expensive. This was historically found to be true both for resource savings and for pollution control, and it fitted nicely with traditional economic theory.

Yet today, not only are there new technologies, but there are also new ways of linking them together so that in principle, *big savings can often be had even more cheaply than small savings*. When a series of linked efficiency technologies are implemented in concert with each other, in the right sequence and manner and proportions (just like the stages in a good recipe), there is a new economic benefit to be reaped from the whole that did not exist with the separate technological parts.

This is a startling contradiction to the everyday wisdom that 'you get what you pay for' – the better, the dearer. How can a slightly more efficient car cost more to build than normal, while a superefficient car costs less to build than normal? There are five main reasons for this. They're explained through detailed energy examples in Chapter 1.

The Purpose of This Book Is Practical Changes

These ideas, at least in broad outline, are not particularly complicated. But they're all new enough to ensure that only a fraction of the world's architects and engineers understand them as yet. Even fewer apply them. Conventional ways of doing things hold practice in a vicelike grip. Most architects and engineers, too, are paid according to what they spend, not what they save, so efficiency can directly reduce their profits by making them work harder for a smaller fee (because their fee is often based, even if indirectly, on a fixed percentage of the project's cost, so if the cost goes down, so does the fee).

Even with proper incentives, it's not easy to apply these new ideas about saving resources. Achieving big savings more cheaply than small savings requires leap frogging, not incrementalism; having the frog get smarter but just sit there in the same old pond isn't good enough. Advanced resource productivity requires integration, not reductionism – thinking about the design challenge as a whole, not as a lot of disjointed little pieces. It therefore fights this century's trend towards narrow specialisation and disintegration. It demands optimisation, not rules of thumb. It requires a new approach to design education and practice. The systems that waste resources today are difficult to design because they're complex, but extremely efficient systems are at least as difficult to design because they're sophisticatedly simple, as many of the examples in Chapters 1 to 3 illustrate.

These essentially cultural barriers to modern resource efficiency are just the tip of a very large iceberg of underlying problems. In attempting to save resources we are faced with a daunting array of practical obstacles that actively *prevent* people and businesses from choosing the best buys first. These include:

- the conventional education of nearly everybody dealing with natural resources, and the often insurmountable costs of replacing conventional personnel with the rare individuals who know better. This 'human factor' may actually be the biggest obstacle and the biggest part of what economists usually call 'transaction costs', the costs of overcoming inertia by taking action to change the way things are;
- other transaction costs related to the massive interests some capital owners have in preserving existing structures – and more inertia from customers who may simply be ignorant about what levels of resource efficiency they could demand;
- discriminatory financial criteria that often make efficiency jump over a tenfold higher hurdle than resource supply (for example, the very common insistence that an energy-saving measure should repay its investment in a year or two, while power plants are given 10–20 years to pay back);
- split incentives between one person who might buy the efficiency and another who would then reap its benefits (such as landlords and tenants, or home and equipment builders and their buyers);
- prices that say little about, and may severely distort, actual costs to society, let alone costs to the environment and to future generations;
- the greater ease and convenience of organising and financing one huge project than a million little ones;
- obsolete regulations that specifically discourage or outlaw

efficiency – ranging from prohibiting taxi drivers who take a passenger into another territory from picking anyone up on the way back, to letting manufacturers' lorries carry only their own goods, to restricting buildings' window areas even when the windows are of a newer kind whose increased area always saves energy, to preferential haulage tariffs that give virgin materials a cost advantage over otherwise identical recycled materials; and

- the almost universal practice of regulating electric, gas, water and other utilities so that they are rewarded for increasing the use, and sometimes even penalised for increasing the efficiency, of resources (an unfortunate side-effect of the restructuring of the British electricity system).

All these obstacles can be overcome, but only by persistent and detailed attention to the real world issues described in Chapters 4 to 7. It requires rewards for saving resources, not for wasting them; procedures for choosing the best buys first, then for actually buying them; and competition in saving resources, not only in wasting them. None of these transformations will be quick or easy; but not making them commits us to solving problems that are far more difficult.

Moreover, the human resources problem mentioned may actually be easier to overcome than we had at first supposed. In places like China, Russia, India and Brazil, the exciting potential of unlocking the brainpower of over 2 thousand million people who were previously excluded from contributing to solutions – as most of the women in the North and West long were – could lead to dramatic progress. While it's not yet clear how to do this, it is clear from some of our examples below (such as 'Fans, Pumps and Motor Systems', Chapter 1, and 'Curitiba's Surface Underground', Chapter 3) that the benefits for the whole world could be enormous.

In summary, although advanced resource efficiency is not easy, it can theoretically be powerful, cost-effective and widely applicable, and is getting more so in practice. In the mid-1970s, for example, the American engineering–economics debate centred on whether cost-effective energy savings could total around 10 per cent or 30 per cent of total usage. In the mid-1980s, the debate spanned a range from about 50 per cent to 80 per cent – a factor of two to five. In the mid-1990s, some of the best practitioners have been discussing whether, in principle, the potential is nearer 90 per cent or 99 per cent – a factor of 10 to 100. As some of our 50 case-studies show, such savings are already being actually achieved by some skilled practitioners. And as the late Anglo-American economist Kenneth Boulding remarked, 'Whatever exists is possible.'

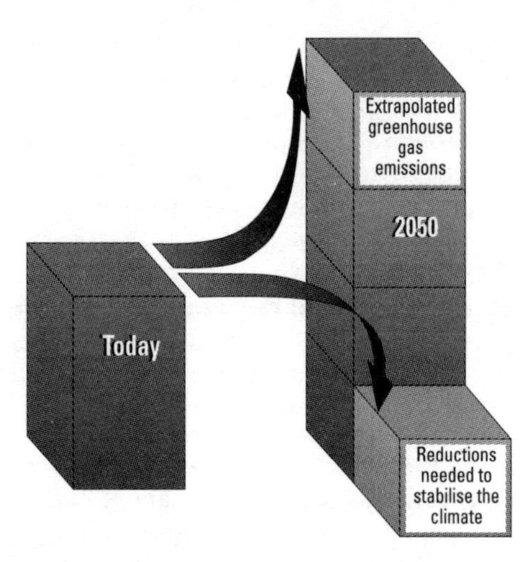

Figure 1: Today's emissions of greenhouse gases could double in the next half-century or so; yet climate researchers think those emissions should be cut in half (Chapter 8). Factor Four efficiency gains can enable us to do twice the work with half the energy – powering prosperity, yet protecting the earth's climate.

Despite the exciting opportunities in the efficiency revolution, we should also take care regarding efficiency's potential to reinforce undesirable patterns. More efficient cars may be driven more and allow for vastly expanding fleets. Saved water may permit further sprawl into deserts. Resource efficiency in general may support unmitigated population increases for an extended timespan. The economic boost from saving resources could thus erode the savings' benefits, if not channelled into a different pattern of development that encourages the substitution of people for physical resources. Chapters 12 to 14 return to this theme of how to harness resource efficiency, not as a tool for doing in a leaner and meaner way things that ought not to be done at all, but rather for humane and worthy goals that meet global needs.

Moreover, we should be realistic about the prevailing incentive structure governing the movements of investment capital, which always has a preference for highest risk-adjusted ROI (return on investment). And we may discover that even highly profitable investments in efficiency will not necessarily be competitive in the capital markets when compared with traditional investments, say, in Indonesian or Zaïrean resource mining or Chinese industrialisation at very poor resource efficiency.

Even faced with all these obstacles and problems, we certainly don't become gloomy. Market conditions and public attention can be influenced. Enlightened consumers can produce strong additional signals in favour of efficiency – and in favour of the obligatory transparency of product labels telling the history of resource use in production (and later in disposal). It is the right of capital owners and of democratic majorities to ask for full information and for level playing field conditions for the efficiency revolution. Chapters 4 to 7 highlight some of our strategic ideas on these lines.

Finally, Part IV considers a brighter civilisation in language that goes beyond technology and quantitative targets. Unavoidably, our economic policies must transcend misleading indicators such as the GDP (Gross Domestic Product), which reflects turnover, not wellbeing. The informal sector, still vital and important in many developing countries, deserves to be rediscovered by our Northern economists. Simplistic views of the benefits of free trade will also need some thoughtful reconsideration.

Part I
Fifty Examples of Quadrupling Resource Productivity

Chapter 1
Twenty Examples of Revolutionising Energy Productivity

People used to talk about 'saving energy'. The phrase had a moralistic connotation. Father would admonish his children to switch off lights when leaving a room and never to let motors or appliances run when not needed. After all, besides costing money, wastefulness was a sin. When a demand for environmental protection became widespread, the reaction on the part of governments, electric power suppliers and some environmental leaders was not particularly imaginative: You (childish and demanding folks) can get as much environmental protection as you want if you are prepared to reduce your demands radically. The simplistic notion of saving energy by voluntarily making do with less allowed leaders to avoid really grappling with the energy issue in a creative way.

In recent years, a new expression has emerged: the 'rational use of energy'. Use of this term enhances the speaker's reputation by suggesting an expertise in energy matters. So we may hesitate to reject this term, yet we aren't happy with it either. It sounds so bureaucratic, complicated and defensive. It doesn't convey any enjoyment and is not straightforward in talking about the relationship between energy use and technological progress.

Technological progress is what this book is all about. Or, rather, it is about redirecting technological progress. Thus we prefer to speak of 'energy productivity'.

Taken by itself, and depending on your circumstances, the term 'productivity' can have positive or negative connotations. These mixed connotations spring from the disservice of economists who have narrowed down the term to mean only labour productivity. In the past, labour productivity was a good thing meaning prosperity. Today, labour productivity is inevitably associated with the threat of unemployment.

Energy productivity, on the other hand, is something everybody can greet with joy. Virtually nobody stands to lose by it.

This chapter is about increasing energy productivity by a factor of four. The expressions 'energy savings' or 'rational use of energy' are simply inadequate to convey the appropriate sense of joyous attack on our prevalent technological dinosaurs. The concept of 'energy productivity' is more equal to the task.

By using Factor Four as a standard, we might at first appear to be excluding much of the manufacturing world: smelting to produce aluminium metal cannot, given the laws of thermodynamics, be made four times more energy-efficient. The same holds for producing chlorine, cement, glass and some other basic materials. But we need not give up on the Factor Four potential of these materials. Aluminium and glass are superbly recyclable, and such recycling would save most of the energy otherwise required to produce them from virgin materials instead. For some end uses, some materials can be substituted for others, with no damage to the manufacturing sector; or materials can be used more judiciously. On a life-cycle basis a factor of four in energy productivity should therefore be available for most end-use services involving metals or glass.

In this book, however, we concentrate on examples with a straightforward potential of multiplying energy efficiency by four or more. We begin with an example that has overwhelming importance for the world's energy balance.

1.1 HYPERCARS: ACROSS THE US ON ONE TANK OF FUEL*

From 1973 to 1986, the average new US-made car became twice as efficient, from 13 to 27 miles per US gallon (17.8–8.7 l/100 km).

* The work summarised in this section is described by numerous papers, ranging from popular to highly technical, available from The Hypercar Centre at Rocky Mountain Institute. A current listing of these and other RMI publications, and the full text of some, can be found at the Institute's World Wide Web site, http://www.rmi.org, and specific enquiries may be directed to hypercar@rmi.org.

About 4 per cent of the savings came from making cars smaller inside, 96 per cent from making them lighter and better; 36 per cent came just from cutting out the most obviously excessive weight. Since then, however, fuel efficiency has risen by only another 10 per cent or so, and as recently as mid-1991, automakers were claiming that by the end of this century only another 5–10 per cent would be feasible without exorbitant cost or sluggish performance.

Can't We Do Better?

The modesty of this claim seemed odd for two reasons. First, many improvements that were used in some mass-produced and well-selling cars were still not being included in many others. Analysts found that the complete adoption of just 17 such widely accepted measures could save another 35 per cent of the fuel used by, say, the average 1987 new car without changing its size, ride or acceleration in any way. The list emphasised such commonplace options as front-wheel drive, four valves per cylinder, overhead cams and five-speed overdrive transmission. It didn't even include some blindingly obvious items such as retracting brake calipers (as motorcycle brakes do) so that they wouldn't drag on the disc and try to stop the car whilst the driver was trying to make it go. And on a conservative estimate, this improvement to 44 mpg (5.36 l/100 km) would cost only 53 cents per saved gallon (14 cents per litre) – less than half today's lowest-ever petrol price in America, where petrol is priced lower than bottled water.

Even as automakers were challenging those analyses, Honda underscored the point by releasing its 1992 VX subcompact, which saved even more – 56 per cent, to 51 mpg or 4.62 l/100 km – and even more cheaply (the bigger saving still cost only 69 cents per gallon or 18 cents per litre). In fact, that Honda car was 16 per cent more efficient than the US National Research Council's 'lower-confidence' estimate, issued several months later, of what will be technically feasible for a subcompact car in 2006!

If that overtaking of prediction by events felt like a time warp, the second reason to think we could do better was positively *déjà vu*. Whatever exists is possible. But almost unnoticed, automakers in the mid-1980s had already created about a dozen concept cars that combined excellent but fairly conventional components in conventional ways to demonstrate doubled or tripled fuel economy. These four- to five-passenger cars achieved 67–138 mpg (1.7–3.5 l/100 km) with normal, and often better-than-normal, safety, emissions and performance. At least two versions, from Volvo and Peugeot, would reportedly cost about the same to mass-produce as today's cars. However, American automakers paid little attention because the concept cars were generally made in Europe or Japan, and so could be assumed to be somehow too different.

By mid–1991, building on these foundations, a far more radical notion was taking shape at Rocky Mountain Institute (RMI). Why not redesign the car from scratch? Why not rethink it from the wheels up, making it radically simpler? Einstein said that 'everything should be made as simple as possible – but not simpler.' Cars, however, had gradually become incredibly baroque, piling one add-on gadget atop another to solve problems that better design should have prevented in the first place.

BACK TO BASICS

Examining the basic physics of cars led to a startling conclusion: the extraordinarily able engineers in Detroit, Wolfsburg, Cowley and Osaka had become so specialised that they knew almost everything about almost nothing; so specialised that scarcely any of them could design a whole car by themselves. Crucial integration between design elements was getting lost. There was too much thinking about little pieces, too little about the car as a system. The industry had lost sight of whole-system engineering with meticulous attention to detail – engineering that is extremely simple, and therefore difficult.

In fact, the auto industry was designing the whole car backwards. After decades of devoted effort, about 80–85 per cent of the fuel energy was getting lost before it could get to the wheels, and only about 1 per cent ended up moving the driver! Why? Because the car was made of steel, which was heavy, and accelerating something so heavy needed an engine so big that it was almost idling most of the time, using such a tiny fraction of its power that its engine became only half as efficient. The industry's answer was to put ever more effort and complexity into wringing slightly more efficiency from the engine and transmission (the 'driveline'). Impressive progress was and continues to be made, but the savings are small and the required effort immense.

But look at the car the other way round. What happens to the 15–20 per cent of fuel energy that does sneak through to the wheels? In level urban driving, about a third heats the air that the car pushes aside (rising to 60–70 per cent at highway speeds), a third heats the tyres and road, and a third heats the brakes. But

Figure 2 Towards the 'hypercar'. A General Motors team built two four-passenger Ultralite concept cars in a mere 100 days. (© General Motors Corporation)

every unit of energy saved from those three fates would in turn save about five to seven units of fuel energy that would no longer have to be fed into the engine in order to deliver that unit of energy to the wheels! Thus, rather than focusing on how to sweat out the next tenth of a per cent of losses in the driveline, designers should treat its inherent inefficiency as a way to multiply energy saved by making the car fundamentally more efficient.

The Ultralight Strategy

Ways to do this were not hard to find. Using ultrastrong yet crash-worthy materials, chiefly advanced composites, could make the car about three times lighter – as little as 1000 lb (473 kg), ready to receive four to five passengers. Better design could make its sleek, moulded shape two to six times more slippery aerodynamically. Better tyres, with less weight pushing down on them, could cut tyre losses three- to fivefold. The car would be designed less like a tank and more like an aeroplane.

This 'ultralight' strategy had already been demonstrated. Indeed, at the end of 1991, General Motors showed a four-passenger carbon-fibre Ultralite concept car with doubled efficiency, excellent safety and cleanliness, great comfort and styling, and very sporty performance (0–100 km/h in 8 seconds, 0–60 mph in 7.8 seconds) – comparable to the acceleration of a 12-cylinder BMW, but with less engine (111 hp, 83 kW) than a Honda Civic. Fifty GM experts had built two Ultralites in a mere 100 days.

This and other experiments showed how very light, slippery construction could make a very attractive car 2–2.5 times more efficient than normal.

HYBRID-ELECTRIC DRIVE

Meanwhile, other experiments, mainly in Europe, had shown that a 'hybrid' electric propulsion system could boost efficiency by 30–50 per cent, partly by electronically recovering up to 70 per cent of the braking energy, temporarily storing it and then reusing it for hill climbing and acceleration. The car would get its energy by burning any convenient liquid or gaseous fuel in a tiny onboard power plant of any kind (engine, gas turbine, fuel cell etc.). Fuel is a better way to carry stored energy than electric batteries, which have less than 1 per cent as much useful energy per pound. That is why battery cars, as Dutch car expert Dr P. D. van der Koogh says, are cars 'that carry mainly batteries, but not very far and not very fast – otherwise they would need to carry even more batteries.'

As RMI's analysts examined the state of the art, they noticed something extraordinary: artfully combining the ultralight and hybrid strategies could raise efficiency not two- to threefold, as one might expect, but about fivefold. It was like finding an equation that said two plus one equals five. Soon, however, the main reasons for this magical synergy were understood:

- weight savings snowball, because the lighter the car becomes, the more components become not only smaller but also unnecessary;
- this 'compounding' of weight savings is even faster with hybrid drive;
- when the ultralight strategy has nearly eliminated the losses of energy that can't be recovered (to heat the air, tyres and road), the only other place the wheel power can go is into braking, and the 'regenerative' electronic braking gets most of that back; and
- those savings of wheel power are then multiplied two- to threefold by avoiding driveline losses to deliver that energy to the wheels.

Thus, if the GM Ultralite were equipped with a hybrid-electric drive, instead of a conventional engine and transaxle, its efficiency would rise not twofold but about four- to sixfold, to approximately 110–190 mpg (1.2–2.1 l/100 km), even before optimisation. RMI researchers soon found ways to improve an attractive family car to several hundred mpg – ultimately enough to cross the US on one tank of fuel (0.8–1.6 l/100 km). And to their astonishment, the car could be so much simpler, and its manufacturing so much easier than stamping, welding and painting steel, that it could end up costing about the same as today's cars – perhaps even less.

The Idea Spreads

In autumn 1993, ISATA, the biggest European car-technology conference, awarded this concept its Nissan Prize as one of the top 3 of over 800 papers presented. Automakers started paying careful attention. The concept was nominated for three US design awards. Press coverage increased. In April 1994, a University of Western Washington student team had a two-passenger Corvette-sized light hybrid car tested by the US Department of Energy in Los Angeles traffic. It got the equivalent of 202 mpg (1.16 l/100 km). In autumn 1994, RMI's research director chaired a 3-day international conference in Aachen on his ultralight-hybrid concept, now called the 'hypercar'. A small Swiss firm, ESORO, displayed a four-passenger light hybrid car rated at about 100 mpg-equivalent (2.4 l/100 km). Distinguished practitioners brought much exciting news, including a fourfold drop in the price of carbon fibre in the past 2 years – to levels that could undercut steel as an auto-body material at any production volume.

By late 1996, more than 25 established and aspiring automakers, of all sizes and many nationalities, had become eager to bring hypercars to market. Some were already making sizeable commitments of resources, totalling around $2 thousand million, to achieve this before their competitors do. The potential of hypercars to cut product cycle times, tooling costs, body parts counts, and assembly labour and space up to about tenfold could give a decisive competitive advantage to companies that get to market first.

Nor are governments standing by. President Clinton's Partnership for a New Generation of Vehicles – the 1993 agreement with the Big Three US automakers to try to develop a tripled-efficiency car within ten years – is providing much helpful support, even though the same and other companies' proprietary programmes will do much better still. California regulators are even expected in 1997 to qualify hypercars as 'equivalent zero-emission vehicles' because they emit even less pollution than do the power plants needed to recharge battery-electric cars – an extra incentive to bring hypercars to market by 2003, when 10 per cent of new cars sold in California must be 'zero-emission'.

READY OR NOT, HERE IT COMES

Today's cars are wonderfully sophisticated and highly evolved – the highest expression of the Iron Age. But many experts believe they are about to be swept away by what may amount to the biggest change in industrial structure since the microchip. And like computer manufacturing, this could happen anywhere in the world, with relatively little capital and with astonishing speed. Likely results: virtual elimination of urban smog; even more cars driving even greater distances (thus making even more urgent the basic transport reforms described in Chapter 6.3); and a more or less permanent crash in the world oil price as hypercars and similarly designed heavier vehicles discover a 'nega-OPEC [Organization of Petroleum Exporting Countries]' – a very attractive and economical way to save more oil than OPEC now extracts.

This may happen very quickly. Two of America's leading experts on efficient cars – Paul MacCready (inventor of the Sunraycer solar car, Gossamer Condor human-powered aeroplane, Impact battery car and many other unique vehicles) and Robert Cumberford (senior correspondent of *Automobile* magazine) – are both saying that by 2005 most cars in the showroom will have electric propulsion, and most of those will be hybrids. They share a rapidly growing belief among car experts that ultralight hybrids, providing the advantages of electric propulsion without the disadvantages of batteries, are the way of the future, and that the future is very near.

Best of all, most people will buy hypercars not because they save 80–95 per cent of the fuel and cut 90–99+ per cent of the smog, but rather because they're superior cars – the same reason one now buys compact discs instead of vinyl gramophone records.

1.2 ROCKY MOUNTAIN INSTITUTE HEADQUARTERS*

In the Rocky Mountains of Western Colorado, 16 miles (25 km) west of Aspen at 7,100 feet (2,200 m) above sea level, is a passive-solar banana farm.

This is not exactly good banana-growing country. The outdoor temperature can fall as low as –47°F (–44°C). The growing season

* For further details see: A B Lovins, *Visitor's Guide*, RMI Publication H-1, 3d ed, 1991, 24 pp.

between hard frosts is 52 days, and mild frost can occur on any day of the year; it recently came on the Fourth of July, violating the normal rule that there are two seasons, winter and July. It is often sunny, but not reliably sunny: midwinter cloud has been continuous for as long as 39 days, and during a recent December and January, the sun was glimpsed on a total of only 7 days.

Nonetheless, at this January writing, three banana crops, one sprouted on the winter solstice, are ripening nicely as it blizzards outside. Two big green iguanas offer Advanced Lizarding lessons to qualified students. Oranges ripen, a waterfall burbles, catfish frolic, and lifelike orangutan dolls, brachiating around the bookshelves, are suspected of coming alive at night – how else to account for the missing bananas? As the days get longer in March and April, the whole jungle will become lush – avocados, mangoes, grapes, papayas, loquats, passionfruit. One walks in out of a snowstorm to the scent of jasmine and bougainvillaea (see Plate 1).

Yet there is no conventional heating system, because none is needed – or cost-effective. Instead, two small wood stoves, run occasionally for backup or aesthetics, provide 1 per cent as much heat as a normal house in this area would need, but the other 99 per cent is 'passive solar'. Even on cloudy days, solar heat is captured through 'superwindows' (see p 19) that insulate as well as 6, or in the latest models 12, sheets of glass: the clear, colourless windows let in three-quarters of the visible light and half the total solar energy, but let scarcely any heat escape. Foam insulation inside the 16-inch (40-cm) stone walls, and in the roof, provides at least twice the normal resistance to heat loss (R–40/k–0.14 walls, R–80+/k<0.07 roof). Air leaks are virtually eliminated, but there is plenty of fresh air – and it's preheated by heat exchangers that recover three-quarters of the heat normally lost in stale air leaving the house.

How much extra did all that heat-trapping cost? Less than nothing: its extra cost was less than the construction cost saved by eliminating the furnace and ductwork.

That leftover money, plus some more totalling $1.50/sq ft ($16/sq m), was used to save half the water, 99 per cent of the water-heating energy (thanks to efficiency and to passive and active solar systems), and 90 per cent of the household electricity. At a tariff of $0.07/kWh (kilowatt-hour), the household electric bill is about $5 a month – before taking credit for the even greater solar production, for which the utility sends a small cheque every quarter.

Daylight, flooding in from all directions, provides 95 per cent of the light needed; superefficient lights save three-quarters of the energy needed for the rest. Controls dim lights according to how much daylight is present, or turn them off after one has left the room. The refrigerator uses only 8 percent, and the freezer 15 per cent, of the normal amount of electricity because they're superinsulated, specially designed and in the case of the refrigerator cooled half the year by a passive 'heat pipe' connected to an outdoor metal fin. The dryer gets its heat from a solar 'clerestory' or lightshaft. The washing machine is a novel top-loading horizontal-axis design that saves about two-thirds of the water and energy and three-quarters of the soap, gets the clothes cleaner and helps them last longer. Even the conventional propane cookstove saves energy with double-walled Swiss pots and a British teakettle, whose heat-trapping design saves a third of the fuel and time needed to boil water. Outside, a superinsulated, passive-solar, photovoltaic 'crittery' helps the pigs gain more weight and the hens lay more eggs because they needn't spend so much energy keeping warm.

Thus to save 99 per cent of the space- and water-heating energy, 90 per cent of the household electricity and 50 per cent of the water, the total extra cost was $1.50/sq ft x 4,000 sq ft ($16/sq m x 372 sq m), or about $6,000 – about 1 per cent of the total project cost in an area with twice the national average building costs. Compared with normal local houses the same size, the energy savings are at least $7,100 a year. The extra cost thus repaid itself in 10 months, after which the saving accumulates at a rate averaging $19 a day, equivalent to a 1.3-barrel-per-day oil well, or about enough to support an intern (a junior researcher). Of course, 10 months is a long time to wait, but that was all done with the best commercially available 1982 technology. Today one can do much better. For example, the windows are now cheaper, but insulate twice as well.

Having paid for themselves in the first 10 months, the energy savings will go on to pay for the entire building in about 40 years. (The building should last at least ten times longer than that; it was built for the archaeologists, who will doubtless conclude from its southern orientation and from the unusual shape of the curving stone walls that it must have been a temple to some primitive solar cult. But it needn't look like that in order to work: it can have any desired shape and style, and be adapted to virtually any climate and culture, yet still save similar amounts of energy and money.) And during that 40 years, just the building's electrical savings will avoid having to burn about enough coal at a power plant to fill up the building twice over. The refrigerator alone saves its own interior volume in coal every year. And the beer stays just as cold.

More than 40,000 people have already visited the building; it has been featured in magazines from *Geo* to *Newsweek*, and in television shows all over the world. Some come to see its technologies and how they're integrated. Some come to see what it's like to combine a house, indoor farm and 20-desk research centre under one roof. (It's nice to commute to work only 10m across the jungle: someone suggested we install vines and swing to work.) But most remark on what may be the building's most important feature: that it helps its occupants feel and work better.

Why does a group sitting around the table stay alert and cheerful all day, when the same group, put in an ordinary office, can get sleepy and irritable in half an hour? Because, we suspect, of the restful curves; the natural light; the healthful indoor air; the low air temperature, high radiant temperature and high humidity (far healthier than hot, dry air); the sound of the waterfall (tuned approximately to the brain's alpha rhythm to be more restful); the lack of mechanical noise, because there are no mechanical systems; the virtual absence of electromagnetic fields; the sound and smell and oxygen and ions (and sometimes taste) of the green plants in the central jungle, open and visible from everywhere. Perhaps there are other things we don't yet understand, but that seems a good list to start with.

Buildings should be net exporters of energy, food and beauty. RMI's headquarters is one of the first highly integrated 'green' buildings, and still one of the best. Many of its details could, we expect, be done much better today, but its basic principles and its level of design integration remain sound and exciting.

1.3 The Darmstadt 'Passivhaus'

In 1983 Sweden adopted a thermal insulation standard making 50–60 kWh/sq m-y the permissible maximum heat loss for houses (the so-called SBN 80, 1983). Average homes in Germany, by contrast, typically show heat losses of 200 kWh/sq m-y. The Factor Four target for Germany would therefore nearly be reached just by adopting the Swedish building standards for all buildings, including older ones. Yet the amended 1995 German standard (*Wärmeschutzverordnung*) requires a mere 20 per cent reduction by 2000 for new buildings.

There are, however, pioneer homes showing that the Swedish building standard can be improved upon considerably. One of the best-known examples is the 'Passivhaus' built in Darmstadt, some 50 km south of Frankfurt. Plate 2 shows the Passivhaus: a very normal looking, inconspicuous building.

'Passiv' refers to the systematic use of passive solar energy and almost no active heating. The Passivhaus shows auxiliary heat demand below 15 kWh/sq m-y (Feist and Klien, 1994). This remarkable value is achieved mostly by using very efficient insulation of both walls and windows.

The house feels at least as solid and durable as do normal German houses. But unlike them, it has an agreeably even temperature. Freedom from mechanical noise (because there is no furnace and almost no mechanical equipment) and street noise (via the sound-suppressing superwindows and thick insulation) create a refreshing silence. And far from being gloomy or stuffy, the house is splendidly full of light and air. Anyone coming into the house feels at once that it provides a superior standard of comfort and restfulness – a cozy feeling of security from the harshness of the outside world, yet also a feeling of connectedness with nature because the large windows bring the green world indoors.

The house uses only 10 per cent of the normal amount of energy for space heating and 25 per cent of the normal amount of electricity. Indeed, the total energy used by the house is scarcely larger than the energy a normal German house uses just for appliances. The space-heating requirements are so small that they are readily met by a slight use of hot water from the gas-fired, superefficient water heater that one must have anyway for domestic hot water: no separate furnace is required.

The windows use somewhat older technology, insulating at the centre of the glass as well as would about eight sheets of ordinary glass (R–8, *k*–0.7). The best windows now insulate about 50 per cent better still, and if used here would about eliminate the last 5 per cent of the original space-heating load. To achieve this complete elimination of space heating, however, would also require an important innovation pioneered and recently installed in the Darmstadt Passivhaus: an overlay of foam insulation that forms a complete cap over the entire window frame and runs 3 cm onto the edges of the glass itself, both inside and out. This window-frame version of a tea cozy eliminates the normal loss of heat flowing out through the window frame, making the edges of the glass insulate just as well as the middle. The system could easily become a mass-produced product, with an attractive plastic surface cap glued to foam beneath, suitable for installation in both new and existing houses and other buildings.

Another important innovation is the 'tempering' of incoming fresh air by flowing it first through a plastic pipe buried 3–4 m underground. Even in midwinter, the ground at that depth is warm enough to heat the cold outside air to at least 46°F (8°C). The prewarmed air then enters a heat exchanger that warms it a further 70 per cent of the way to the temperature of the warm, stale air leaving the house. Thus the virtually airtight house gets lots of fresh air all the time, but loses hardly any energy in the process. The airflow can be separately controlled to different parts of the house, and the more people are in a given part, the more fresh air they will get, because as they breathe, a sensor detects the resulting carbon dioxide (CO_2) and adjusts the quiet ventilation fan accordingly.

The flow of heat in and out of the house is as precisely measured and thoroughly understood as for perhaps any building in the world. To achieve this degree of agreement between theory and experiment, it was necessary to account for terms normally too tiny to worry about: the exact placement of sensors within the walls to within a fraction of a millimetre; the loss of heat to cold water that comes into the house, sits inside the toilet tanks and is then discharged when flushing; and even the tendency of the plaster to 'breathe', absorbing and re-evaporating water vapour in different seasons – an effect responsible for about a tenth of the space-heating load.

The costs were higher than those of ordinary buildings. In many regards, a substantial price had to be paid for obtaining pioneer materials and services. The next step was to adapt the concept to standardised and cost-effective construction methods. The breakthrough came in 1996. One of the Darmstadt architects, Folkmer Rasch, designed council houses at competitive prices and with the efficiency standards of the Passivhaus. This earned him the newly-created Schuler Award for resource efficiency. In time for the World Exhibition, Expo 2000 in Hanover, an entire city, the Kronsberg Siedlung, will be completed, with energy efficiency four times higher than new buildings and at no extra cost.

1.4 Hot-Climate Houses in California

The previous two examples described buildings that require only 1–10 per cent the normal amount of space heating in cold and cloudy climates. But what about space cooling in hot climates?

The largest investor-owned American utility company, Pacific Gas and Electric, is operating an experiment called the Advanced Customer Technology Test for Maximum Energy Efficiency – 'ACT^2' for short. Its goal is to determine by careful measurement the most energy that can be saved, cost-effectively to the utility and attractively to its customers, by installing the best-integrated packages of modern technologies. The experiment is guided by a committee of PG&E, RMI, Lawrence Berkeley National Laboratory (the top national laboratory on energy-efficient buildings) and Natural Resources Defence Council (a leading national conservation group).

Cofounded in 1989 by PG&E's research manager, Carl Weinberg, and RMI's research director, Amory Lovins, this 7-year, $18-million effort has now built or retrofitted all of its dozen experimental buildings, and the data are starting to pour in. They broadly confirm the initial hypothesis that about three-quarters of the electricity used in most situations can be cost-effectively saved, whilst providing the same or better services.

A new house in Davis, California, near Sacramento, provided an early test of how far designers could go in a challenging climate.* The 'design temperature', which occurs only some of the time, is 105°F (40°C), but the very hottest days actually reach 113°F (45°C). Extensive irrigation of surrounding farmland and lawns adds substantial summer humidity. Although summer nights are often cool, multiday 'heat storms' with little night-time relief can also occur.

The task was to design an ordinary-looking, completely comfortable, 1,672-gross-square-foot (255-sq-m) tract house with standard layout and amenities, using as little energy as possible. But before any improvement, the 'base case' house already met the strictest energy standards in the US – the most restrictive interpretation of California's 1993 Title 24 code, already about a third more efficient than average US construction. The $249,500 price was also in the typical California midrange.

First the design team at Davis Energy Group eliminated 11 per cent (seven metres) of unnecessary perimeter length, which costs money and wastes energy, by improving the floor plan, but got the same visual interest by manipulating the lines of the roof. Then they put the windows in the right places, improved the window frames and used an engineered wall that saved wood (see 'Wood in Home Building', Chapter 2.20), cut construction costs, and doubled the insulation. Results so far: a 17 per cent energy saving, at a 'mature-market' construction cost – *i.e.* if the house's experimental features were to become widely adopted – nearly $3,500 less than normal. Of that cost reduction, 57 per cent was from the new floor plan's smaller perimeter, and the rest would apply to practically any new house.

Next, the designers introduced a swarm of small improvements to the shell, lights, appliances, hot-water system and windows,

* Detailed publications are available from ACT^2, PG&E Research, 2303 Camino Ramon, Suite 200, San Ramon CA 94583, tel 510-866-5573. See also Lovins (1995).

raising the total energy savings to 60 per cent and costing nearly $1,900. The only unusual measure was using the refrigerator's waste heat to preheat hot water: this saved water-heating energy, made the refrigerator more efficient and made it cool rather than heat the house. Some savings were small but instructive, such as the exhaust fans, where careful shopping saved 80 per cent at no extra cost. Their normal efficiency in North American houses, it seems, is an abysmal 1–3 per cent: they're really little electric heaters that happen to use a tiny fraction of their energy to move air. Along the way, the doubled wall and roof insulation and more efficient windows eliminated the $2,050 furnace and its associated ducts and equipment. In its place, a little hot water from the 94 per cent-efficient gas-fired water heater could be run through a $2,400 radiant slab coil on the coldest nights.

Yet one-third of the original 3-ton* (12.3 thermal kW) air-conditioner remained, and the costliest improvements were already up to the cost-effectiveness limit: no measure could cost more than the expected long-run price of the electricity they saved (6 cents per kilowatt-hour). What more could be done?

Happily, the designers had kept aside a special basket of measures called 'potential cooling elimination package'. In it they had put every energy-saving measure they had rejected because it didn't save enough energy to pay for itself, but that *also reduced the need for space cooling* and had not yet been given proper economic credit for doing so. Sure enough, when seven such measures costing $2,600 were added to the design, they more than eliminated $1,500 worth of air conditioner and ductwork (plus its $800 of future upkeep costs), thereby becoming cost-effective. Thus *big savings became cheaper than smaller ones*.

The designers expected, and the first summer's occupancy confirmed, that no backup cooling would be needed. Insulation and elementary superwindows kept unwanted heat out; efficient lights and appliances released little heat inside; thermal mass (double drywall and, in the central zone, a ceramic-tile floor) stored coolness to 'ride through' daily temperature peaks. If necessary, cool city water about to be used in the household could also be run through the slab coil for free radiant cooling. This proved impractical, because the water main turned out to be buried too shallowly, making the city water unexpectedly warm; but fortunately the calculations were correct, so it was not needed anyway.

---

* A 'ton' of cooling measures the rate of providing cooling. It's the rate provided by melting one American ton (2,000 lb or 907 kg) of ice over a 24-hour period, and equals 3.52 thermal kW.

Computer simulations predicted that the house would use 53 per cent less electricity, 71 per cent less on-peak electricity and 69 per cent less natural gas than the already very efficient base-case house. But this assumes no improvement in the small appliances that used one-third of the initial electricity. Taking that out of the comparison would raise the energy savings to an average of 80 per cent for all energy or 79 per cent for electricity alone: 78 per cent for space heating, 79 per cent for water heating, 80 per cent for refrigeration, 66 per cent for lighting, 100 per cent for space cooling, and 92 per cent for space cooling and ventilation together. Yet in a mature market, the Davis house, complete with its 20 kinds of energy-saving measures, would cost about $1,800 less than normal to construct and $1,600 less than normal to maintain.

Early monitoring results suggest that the house is working better than designed. (Actual savings were decreased somewhat by last-minute changes in the design, such as the occupants' wanting a different kind of refrigerator; but accounting for those changes, measured savings agree well with predictions.) The occupants, who moved in in December 1993, are satisfied with its comfort, even in a severe heat storm. Since California's Title 24 energy standard is supposed to include everything that is practical and cost-effective, the Davis house may require the standard to be fundamentally rewritten.

Another ACT^2 project completed a few months later, a comparable tract house in even hotter Stanford Ranch, California, is saving even more energy. It's largely passive: better orientation, insulation, windows, light-coloured walls and roof, and double drywall reduce the cooling load by 44 per cent. It also uses a direct evaporative cooler, run only at night, which serves as a whole-house fan and cools water that circulates through under-floor tubing, raising the total cooling-energy saving to 86 per cent. The Stanford Ranch house costs about $1,000, or 0.4 per cent, more than usual to build, but its saved replacement and maintenance costs offset that slightly higher construction cost, so the net cost of roughly quadrupling the efficiency of all the major energy uses is zero.

A third ACT^2 project, an ordinary 15-year-old one-storey house, was retrofitted in spring 1994. It's in a hot area in Stockton, California, where people normally run air conditioners from June through September. Details of the house and special requirements of the occupants also restricted how much could be done. Nonetheless, the retrofit is expected to save 64 per cent of the house's total electricity and 60 per cent of its natural gas (not counting some further potential savings in small electric appliances) at a mature-market cost of $5,500, including future effects on replacement and maintenance costs. Simple improvements to the air conditioner – more efficient blower motors and fans, and an evaporative precooler – contribute to an expected 76 per cent reduction in cooling energy. In addition, heating energy is expected to drop by 59 per cent, major appliances by 63 per cent, lighting and small appliances by 76 per cent, and pumping energy for a swimming-pool and spa by 76 per cent. If the occupants had usage patterns more typical of American families, several of these figures would be even higher.

Although all three ACT^2 houses are in hot climates, none is in a muggy tropical climate. Yet comparable savings have been achieved there too. Thus neither heat nor humidity is an obstacle to Factor Four savings with excellent comfort and profitability.

1.5 Superwindows and Large-Office Retrofits

Superwindows use invisibly transparent high-tech films to sort out visible from infrared (heat) radiation. The visible light goes through; the infrared is reflected away. Superwindows now come in hundreds of thousands of different 'flavours', each suited to a specific climate, building and direction of orientation. Sophisticated designers 'tune' a building's windows by using a different kind on each side – letting in a lot of light and heat on the poleward side, minimising heat gain on the sunny afternoon side and so forth. All these different kinds of superwindows look the same, but their infrared properties differ. By independently controlling the flow of heat and light in and out of the building in each direction, the designer can improve comfort, greatly reduce the need for heating and cooling equipment and for energy to run them, and thus reduce both construction cost and operating cost.

Superwindows started to enter the US market only in the early 1980s; RMI's headquarters may have been the first commercial project to combine superwindows' spectrally 'selective' thin films with an insulating filling of heavy gas. Improved multifilm versions, tested at RMI in the early 1980s, came on the market a few years later. Heat Mirror suspended plastic film coated on both sides, packing more performance into a thinner window, went into production only in November 1993. And European availability of comparable products has lagged by many years, with modern superwindows of reasonable thickness and price starting to appear only in 1993–94.

Essentially all European superwindows are still designed to maximise heat gain in cold, cloudy climates. But cooling equipment is far costlier than heating equipment. This makes it even more valuable, as most US window firms now do, to optimise superwindows to minimise heat gain in hot climates – and even in colder ones, too, if the building is so big that it's self-heated by the lights and equipment inside. That is true of most large, modern office buildings: in Toronto or Stockholm, many buildings must be air-conditioned even at 14°F (–10°C)!

Traditionally, hot-climate windows are either reflective, causing annoying glare outside, or dark and heat absorbing, so that half the heat is reradiated back inside anyway. Both solutions limit the entry not only of unwanted heat but also of desired daylight. The resulting gloominess must then be corrected by electric lights. They use electricity, which ends up as heat inside the building, so you're back almost where you started. But superwindows can be designed to let in desired daylight while largely blocking unwanted heat. Some recent designs, using slightly aqua- or green-tinted glass, let in visible light about twice as well as total solar energy. That is theoretically about perfect: you can't do better, because half the solar energy is infrared to start with.

Such hot-climate superwindows can also keep you warm in the winter. Their insulating value comes from their heavy gas filling and from their special films' ability to reflect back into the building the far-infrared rays that try to escape, robbing the occupants of valuable heat. Thus superwindows can be chosen to provide excellent year-round performance in climates with both hot summers and cold winters.

Normally superwindows are assumed to be worthwhile only in new construction. After all, who wants the expense of having to remove and replace an existing window, especially high on a big building? In fact, sometimes the window's extra value isn't worth its cost, especially including labour to install it. But there are important exceptions.

In 1988, RMI studied for then-governor Clinton the potential for saving electricity in Arkansas (Lovins, 1988). Added onto a typical old wooden single-family house, a package of about 20 carefully chosen measures could save 77 per cent of the annual and 83 per cent of the peak electricity (and 60 per cent of the gas just through better insulation, without improving the gas appliances) while paying for itself in only 3 years. The key was to add heat-blocking superwindows right over the existing windows – unshaded single panes of clear glass. This retrofit reduced cooling loads far

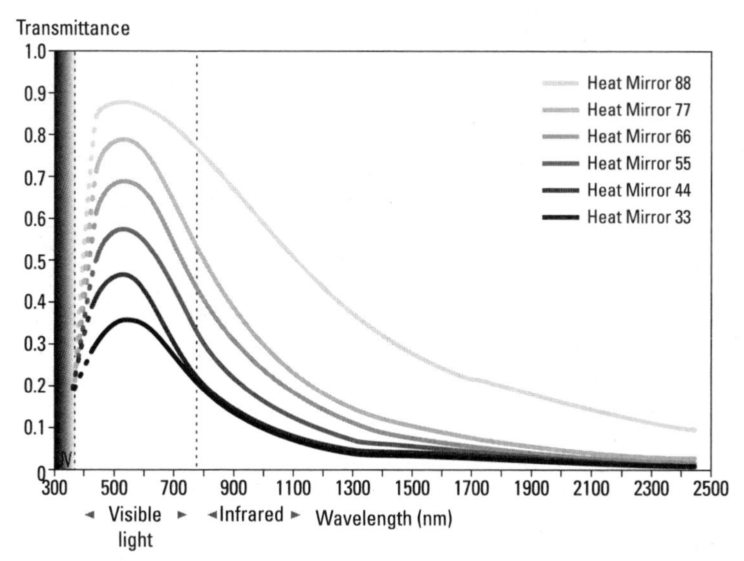

Figure 3 Superwindows let daylight pass while reflecting long heat rays. The graph shows the reflection characteristics for the relevant part of the electromagnetic spectrum.

more than any other measure, and so was largely responsible for making the replacement air conditioner two-thirds smaller. At that smaller size, a unit more than twice as efficient, and dehumidifying three times as well, cost virtually the same, making its even larger electricity savings nearly free. Taking account of all the interactions between the size, cost and performance of the different measures, using the relatively expensive superwindows made the total electrical savings much greater and one-third cheaper.

In 1994, RMI revisited this logic on a much larger scale (Lovins, 1988). A major corporate owner was renovating a 20-year-old, 13-storey, 200,000-sq-ft (18,587-sq-m), all-glass office tower near Chicago. Its construction was standard early-1970s 'curtainwall': big windows were flush-mounted on a steel frame, with transparent 'view glass' alternating with dark-painted 'spandrel glass' that spanned the gap over the steel and concrete structural elements between the floors. Each kind of glass covered half the building's skin area. Since the view glass was double and the spandrel was single and uninsulated, the building shell's average insulating value was less than that of two sheets of glass – grossly inadequate for the severe year-round climate – and leaked air and water to boot. Indoor comfort left much to be desired despite huge heating and cooling systems.

Double-glazing ('insulated-glass') units do have a disadvantage: their edge-seals eventually fail, causing them to cloud up inside. Today's best units have a typical seal life of 23 years; the cheaper kind, 12. Twenty years ago, they weren't even that good. So when RMI found that 8 per cent of the 900 double-glazed units had already failed, its experts suggested a special 'frost test' of the rest. It disclosed that virtually all of them could be expected to fail and need replacement during the next 6 years – a major nuisance for prospective tenants. The owner therefore decided to reglaze the whole building before bringing in the next set of tenants.

The existing glazing, however, was dark double-bronze glass plus a sun-blocking gray film – a combination so dense that only 9 per cent of the daylight could get in, creating a cavelike mood and isolating the occupants from their environment. Moreover, such glazings are so expensive that a superwindow would cost scarcely more to buy, and slightly less to install. Yet the superwindow would insulate three times as well, let in six times as much daylight and block unwanted solar heat so effectively that, together with efficient lights (p 36) and office equipment (p 41), the cooling load could be cut nearly fourfold – from 750 to fewer than 200 tons of air-conditioning.

Then the owner would have an additional opportunity. Air-conditioning equipment normally needs renovation every 20-odd years as moving parts like fans and pumps wear out. In addition, during the 1990s the owner will need to deal with the phaseout of the big chillers' CFC (chlorofluorocarbon) refrigerants. But instead of a normal like-for-like replacement, costing about $800 per ton (plus some more to get rid of the CFCs), the owner could replace the whole space-cooling and air-handling system with a better-designed version nearly four times as efficient (see pp 53–64) and costing perhaps as much as $2,000 per ton. That is a good buy: cost per ton goes up perhaps 2.5-fold, but the number of required tons – the amount of air-conditioning capacity – falls nearly fourfold. That means the construction cost goes down. Then the owner can use the saved air-conditioning renovation cost to pay for the lighting and day-lighting retrofits.

RMI calculated a striking net result. Peak electricity demand – the kind utilities most worry about because it determines how much costly equipment they must build – would fall by 76 per cent. Annual electric use would fall by at least 72 per cent and probably more. Amenity and aesthetics would improve dramatically, making it far easier to recruit and retain tenants: indeed, the building would set an altogether new standard of comfort, quiet and beauty. Operating costs would fall by $1.10/sq ft-y (nearly $12/sq m-y) – some 10–20 times as big as competitive rent differences, giving early adopters a huge market advantage. And balancing out all the construction costs and savings, the whole renovation, saving three-quarters of the energy, would pay for itself in somewhere between –5 and +9 months – that is, more or less immediately.

In the event this project was not implemented; but its principles remain sound, and its analysis used generally accepted tools and field-proven assumptions. Similar retrofits remain a lucrative opportunity for more than 100,000 large curtainwall buildings of similar vintage (old enough, around 20 years, that the windows and mechanicals need renovation) in the US, and probably even more abroad.

And why was such a profitable opportunity not grasped by its owners, who after all had commissioned the study in the first place? Because of the kinds of market failures we'll discuss further in Part II:

- ❑ a lack (at that time) of good ways to share savings between owners and tenants, or to reward designers properly for their extra work in achieving such large savings, or to educate tenants about the mutual benefits of making their lights and office equipment far more efficient;
- ❑ an improper emphasis on wringing cost out of each building component separately rather than out of the building as a whole; and
- ❑ in this case a particularly perverse little detail: that the local leasing office which controlled the property was incentivised to refill the building with tenants as quickly as possible, and reluctant to delay its commissions from those leasing deals long enough to accomplish the retrofit.

In the end, the building was renovated in the old, inefficient way. It then proved too costly and uninviting for the market to accept and, being unleasable, had to be sold at a distressed price – the cost of not grasping the new design opportunities. RMI, however, is undeterred: it is easy enough to work instead with the owner's competitor down the road. Sooner or later, the owner will learn what makes market sense. Competition from more efficient properties can concentrate the mind wonderfully.

1.6 QUEEN'S BUILDING, THE NEW SCHOOL OF ENGINEERING AND MANUFACTURE, DE MONTFORT UNIVERSITY, LEICESTER UK

Officially opened in December of 1993 by Her Majesty, Queen Elizabeth II, this British academic building, designed by Alan Short and Brian Ford of Peake, Short, and Partners, works in impressive harmony with nature (see Plate 3). 'The architects were prepared to take on both the environmental and architectural challenge in the same project rather than ducking out of one to pursue the other. [E]nvironmental responsibility does not mean bland architecture.'

In this passively cooled and ventilated engineering school, students of mechanical engineering study chalkboard diagrams of the chillers it doesn't have, while electrical engineering students study electric lighting design in a naturally daylit space with the lights all off. 'We felt,' said the designers, 'that experiencing the natural rhythm of the world outside helps achieve quiet concentration, and that a school for engineers which achieved significant energy savings could itself be a teaching tool and research vehicle.'

Carefully sited to follow closely the road to the northeast, this addition to De Montfort University's campus leaves much of the site open and blends into the campus with careful landscaping by Livingston Eyre and a planned park on the south side. The original programme for the building design listed three guidelines:

- use traditional labour-intensive construction methods to create jobs for local workers,
- show innovative new concepts that provide an excellent educational environment and challenge conventional architectural practices, and
- use cleaner and greener technology.

Inspired by Trinity Lane, a medieval street in Cambridge, the design is actually a collection of smaller, domestic-scale buildings – supplemented by a series of courtyards that double as outdoor classrooms – so its 110,000 sq ft (10,223 sq m) don't seem overpowering. This charming structure has actually been classified as the first Gothic Revival building in the past hundred years. Resident masons, many of whom needed work in the neighbourhood, created the beautiful polychromatic brickwork, whose wonderful assortment of traditional brick detailing greatly enhances the building's aesthetics.

The Queen's Building is the largest naturally ventilated building in the UK. Its narrow floorplate not only allows daylight to penetrate deeply from two sides, providing nearly all-natural lighting, but also permits almost entirely passive air movement. This occurs both through cross-ventilation and by the rising of warm air up eight large decorative chimneys capped with ornamental metal louvres. Partly government-funded design efforts including fluid-mapping tables, physical models and computer simulations created a passive ventilation design for the auditoria. Occupants can open or close windows to adjust comfort conditions; 60 per cent of the shell area is operable. To ensure acceptable temperatures, an automated management system adjusts dampers, louvres, and heating controls.

Calculated overhangs and heavy masonry walls minimise cooling loads, and the whole design of the building minimises heating and air-conditioning demands. Heating is primarily by passive solar design and by internal gains from extensive engineering equipment and from the thousand workers and students. These heat loads can be quite high – 84 W/sq m from equipment in the electrical lab, and up to 100 W/sq m in the mechanical lab, compared with only 25–32 W/sq m for the equipment and people in typical UK office buildings. Supplementary heating is provided by natural gas.

All of these strategies minimise electricity use. This makes equipment smaller or even eliminates it, saving both energy and capital cost. Only 24 per cent of the capital cost of the Queen's Building was for mechanical and electrical systems, compared with a typical 34–40 per cent. The building also uses only 25–50 per cent of the fuel of an equivalent building. The building's $12 million total cost was only $184/sq ft ($1,980/sq m) finished and completely equipped, or $110/sq ft ($1,184/sq m) unfinished – both extraordinarily low for an engineering school.

1.7 Renovating Masonry Row-Houses

In St Louis, Missouri, as in most American cities and many cities elsewhere, much of the old housing stock consists of street after street of narrow, three-storey row-houses made of stone or brick. Many of these solid, classic old structures in St Louis are run down, even derelict. What can be done with them?

When he was head of the St Louis Energy Office, Jim Sackett

realised that the city's housing future lay in somehow rehabilitating these buildings. But many were in dreadful condition. Not only were the interiors falling to bits – sometimes even destroyed by fire – but sometimes the passage of time had made the walls not quite vertical, the floors not at all level and practically nothing straight. Yet the skilled work needed to make everything plumb, square and level was unaffordable. And even if the money could be found, how could the city's many poor people ever afford to pay the energy bills in these houses, with their thin or nonexistent insulation, in a climate that ranges from very hot and humid in summer to quite cold in winter?

With much trial and error, a successful solution was found. A 'float concrete' floor, somewhat foamy with entrapped air, was first poured into the bottom as a self-levelling foundation. A neighbourhood workshop would then mass-produce prefabricated panels that sandwiched insulating foam between fireproof drywall layers, with the edges designed to lock tightly together. With a simple mounting system adapted to highly imperfect walls, the entire building could be lined with this attractive and superinsulating finish material.

An easy way was then found to cut in the openings for doors and windows. Since superwindows were hardly yet available at that time, Sackett chose a widely available alternative: two double-hung windows, each with an upper and a lower element sliding in adjacent tracks. By opening and closing the inner and outer, upper and lower, sashes in various combinations, the windows could be made to heat, cool, ventilate or insulate the house in various seasons.

The result was as remarkable in its performance as in its simplicity. Without using fancy methods or materials – no superwindows, no air-to-air heat exchangers and so forth – the superinsulation retrofit cut the life-cycle cost of each house in half. Heating needs fell by over 90 per cent; even with no heating at all, in the most bitter winter, the inside temperature could never fall below about 55°F (12°C) because of passive-solar gain through the windows. Cooling needs fell by nearly as much, so that one small window air conditioner could cool and dehumidify the whole house enough for good comfort.

The cost? Less than $2,000 beyond the cost of the basic rehabilitation alone, yet enough to turn a derelict stone shell into an elegant, durable, attractive, safe and affordable home.

1.8 ING Bank Headquarters

In 1978, Nederlandsche Middenstandsbank (NMB) was the number four bank in the Netherlands. Now it's number two and has been renamed ING Bank (International Netherlands Group). This bank has been steadily growing and was strong enough to acquire Barings Bank after its breakdown following derivatives speculation.

How did it come to that? It's a long story, almost as good as a fairy tale, but true. And it has a lot to do with resource efficiency.

In the old days, NMB was 'stodgy and conservative', as one of its managers conceded. Needing both a new image and a new headquarters, the bank's employees voted on a site in the growing area south of Amsterdam. The bank's board of directors wanted a building that was organic, integrating art, natural materials, sunlight, green plants, energy efficiency, low noise and water.

The bank put together a team to design and build it. The team was told to work across disciplines, said Tie Liebe, the bank's real-estate manager (whose help with this case-study is gratefully acknowledged). The integrated team worked for three years on the design, regularly consulting the future users. Construction began in 1983 and was completed in 1987. The result is a highly unusual design: the architect, Anton Alberts, describes its style as 'anthroposophical', a philosophy based on the works of the German educator Rudolf Steiner. Look at it and you see that there are no right angles in the building (see Plate 4).

The building, which accommodates 2,400 employees on some 50,000 sq m (ca. 538,000 sq ft), is broken up into a series of ten slanting, brick-faced, precast-concrete towers. The ground plan is an irregular S-curve, with gardens and courtyards interspersed over the top of some 300,000 sq ft of parking structure and service areas. Restaurants and meeting rooms line the internal street that connects the ten towers. Architectural historian Charles Jencks (1990) described the building as a 'groundscraper', with an 'undulating body ... that hugs the earth.' The high-density residential, office, and retail development surrounding the bank reinforces the image of a medieval castle with its surrounding village.

Like most northern European offices, the complex uses floorplates narrower than the North American norm. Maximum floor depth is small enough that no desk is more than about 23 ft (7 m) from a window, permitting excellent day-lighting. Interior louvres are used to bounce daylight from the top third of the exterior windows onto the ceiling of office spaces. Together with window-lined interior atria that penetrate through the towers to the mezzanine-level internal street, this floor-by-floor sidelighting provides a significant part of the building's total lighting, supplemented by task lighting, custom decorative wall sconces and limited overhead fixtures.

A similar level of concern distinguishes the building's thermal design, permitting largely passive design in the cloudy Dutch climate even though superwindows weren't yet available in Europe, confining the designers to ordinary double glazing. A sheath of insulation separates the brick skin from the precast-concrete structure. The structure itself is used to store heat from simple passive solar measures and from internal gains such as lighting, equipment and people.

Additional heat is supplied through hydronic radiators connected to a 26,420-US gallon (100-cu m) hot-water storage system in the basement. This water is heated by a cogeneration facility located within the structure, and by heat recovery from the lift motors and computer rooms. The NMB building also uses air-to-air heat exchangers to capture heat from outgoing exhaust air and preheat intake air. Like many northern European buildings, the bank is not air-conditioned; instead it relies on the thermal storage capacity of the building fabric, mechanical ventilation, natural ventilation through operable windows and a backup absorption cooling system (chiefly for dehumidification) powered by waste heat from the cogeneration system.

This level of integration between building design, daylighting and energy systems yields impressive results. NMB's old head office building consumed 422,801 Btu/sq ft (4.8 GJ/sq m) of primary energy annually. The new building consumes 35,246 Btu/sq ft (0.4 GJ/sq m) annually – 92 per cent less. In contrast, an adjacent bank constructed simultaneously with the NMB building consumes five times as much energy per square foot, and had roughly the same construction cost (Olivier, 1992). The additional construction costs attributed to NMB's energy systems was around $700,000; however, the annual energy savings are estimated at $2.6 million, for a 3-month payback (Olivier, 1992). Liebe notes that NMB has the 'lowest energy costs in Dutch office buildings, and is one of the lowest in Europe.'

The same level of design integration that is evident in the energy systems can be seen in the incorporation of artwork, plants and water amidst the finishwork's simple natural materials. Circulation spaces throughout the bank are filled with artworks, and not just 'plop art': in many cases the artistry is part of the building fabric.

For example, pieces of coloured metal in the top of the tower atria reflect coloured light down to light sculptures in their base, which then bathe the surrounding stuccoed walls in coloured light. This attention to detailing extends even to the treatment of expansion joints. Where the brass plate covering an expansion joint in a major corridor travels up a wall, it becomes a piece of relief sculpture, recessed into the wall and surrounded by a fan of varicoloured marble and cove lights. Rooftops, courtyards, atria and other interior spaces are landscaped using a variety of garden styles. Cisterns capture rainwater for use in fountains and landscaping. 'Flow form' sculptures, which create a pulsing, gurgling stream from a constant flow of water, are used extensively – even as handrails for multi-storey ramps. (Bank workers in three-piece suits can be observed playing in the handrails.) Beyond their visual appeal, the water features serve to add moisture to the air and to add an acoustic mask to overcome the otherwise noticeable silence of a building that lets little outside noise in and, being largely passive, creates almost none of its own.

Construction costs for the building were roughly $1,500 per square meter (ca. $145 per square foot) including land, structure, landscaping, art, furniture and equipment. That is comparable to or cheaper than other office buildings in Holland. Moreover, absenteeism among bank employees has dropped, which Liebe attributes to the better work environment (Romm and Browning, 1994). The building has also done wonders for NMB's public image. NMB/ING is now seen as a progressive, creative bank; the building is the best known in the country after the Parliament House, and the bank's business has grown dramatically.

1.9 Cutting Danish Appliances' Electricity Use by 74 per cent

About 30–50 per cent of the electricity in most industrialised countries, and 45 per cent of that in Denmark, runs appliances (including household lighting, hot water and ventilation) in homes and in the service sector. Very careful and detailed analysis at the Technical University of Denmark, emphasising major household appliances, has shown how 'the present level of electricity services – such as cooling, cleaning, cooking and provision of clean air – can be maintained while using only 26 per cent of present electricity consumption, if efforts are directed towards development and implementation of efficient technology' (Nørgård, 1989).

The estimated added cost for such efficient appliances averages only 2.5 US cents per saved kilowatt-hour – equivalent to just the fuel cost from a power plant burning $14-per-barrel crude oil. Indeed, most of the savings can be achieved simply by using the best appliances already on the market, whose extra cost averages only 0.6 cents per saved kilowatt-hour. These costs of saved energy might increase a little with an adjustment for the slightly larger amount of space heating required when appliances become more efficient (since as much as two-fifths of their energy contributed useful heat to a typical Danish household), but have also fallen even more because of better technologies since the analysis was done.

How can a country that is already relatively efficient, such as Denmark, quadruple the efficiency of typical household equipment compared with 1988? It isn't so hard after all:

- Better insulation, compressors, refrigerants, heat exchangers and controls had already reduced the annual energy needed by a typical Danish 200-litre refrigerator (with no freezer) from 350 kWh in the 1988 stock to 90 kWh for the best sold in 1988. The Danish analysis further found a clear potential to achieve only 50 kWh with either an advanced motor/compressor design or 'free' cooling from the normally cool outdoors. As described in 'Super-refrigerators', on page 33, this finding turned out to be conservative: the former approach actually achieved 50 kWh in a recent Dutch machine, the latter achieved 38 at RMI, and adding vacuum insulation to the Dutch version would reduce it to about 30 kWh.
- Similarly, the best freezer on the Danish market in 1988 used about 64 per cent less electricity than the average then in use (180 vs. 500 kWh/y for about 250 litres), but a more advanced design could readily achieve only 100 kWh/y, or an 80 per cent reduction. The same 80 per cent reduction is straightforward for a combined refrigerator-freezer.
- The average Danish household clothes-washing machine in 1988, with a capacity of 4 kg (nine pounds), was run about 200 times per year (using about as much time as was used for washing clothes by hand in an earlier generation!). This used about 400 kWh of electricity, including the normal European electric booster to heat cold water coming into the machine. But the best available model used only 240 kWh, and a more advanced design used only 115 – comparable to the best on the US market by 1994. Further ways to reduce this electrical

usage to only 40 kWh/y (substituting non-electric sources of water heating) were clearly in view. The improvements weren't only in the washing machine itself but also in better detergents, some of which can now effectively dissolve fat even in cold water. Some efficient machines also separated, and separately optimised, the processes of soaking (which needs concentrated detergent) and mechanical agitation (which needs more water). Further examples of innovations include sensors that keep adding water and soap until the rinse water comes out clean and nongreasy, then stop – typically reducing energy and soap use by many-fold compared with 'dumb' machines that lack such sensors.

- Dishwashing machines could similarly be reduced from the 1988 average of 500 kWh/y (with four uses per week), or 310 for the best model then available, to 165 kWh/y through better design, or to only 35 if other heat sources were used to heat the water. The main improvements are better motors and pumps, better detergents, thermal insulation and controls.
- Clothes dryers, averaging 440 kWh/y with 130 3.5-kg loads per year in 1988, could be reduced to 350 kWh/y with the best 1988 models, 180 with better ones and only 100 with nonelectric heating. The most obvious improvements include insulation, efficient motors, smarter controls, heat pumps and possibly microwave drying. Major savings are also available from spinning the clothes much faster: some modern washing machines can do this (increasing their motor energy use 19 times less than they reduce drying energy later), yet automatically shake the clothes afterwards to get out the wrinkles.
- Danish electric equipment for household cooking typically used about 700 kWh/y in 1988, but the best available models used only 400, and advanced versions could work even better with 280, even without the obvious step of substituting natural gas. Some improvements in electric cooking are very simple, such as better thermal contact between heating element and pot (often only 30 per cent of the heat gets into the pot today), built-in heating elements, better-insulated oven inserts, insulated pots, and pressure cookers. The research team, for example, built an electronic control that measures the temperature at the bottom of the pot and provides exactly enough heat to reach the desired temperature without overshooting. Even a classic Danish rice-in-milk dessert that normally requires tedious stirring to keep the milk from scorching would come out perfectly with no stirring at all!

- Another important, though often overlooked, appliance is the little pump that circulates hot water from the furnace throughout the house. The standard Danish pump in 1988 used 65 watts, when the actual circulating energy required in the water was only on the order of 1 watt. By 1988, a cheaper 20-watt pump and better controls reduced this 400 kWh/y usage to about 100 kWh/y. A more efficient 5–10-watt model using off-the-shelf technology could clearly reduce this to 50 kWh/y. Naturally, superinsulation and ventilation heat recovery and superwindows could greatly reduce or even eliminate the furnace and hence its circulating pump as well.
- Ventilation – common in larger buildings and being introduced into relatively airtight houses – is often extremely inefficient. Common kitchen and bathroom exhaust fans in North America have been measured to be only 1–3 per cent efficient, although more efficient and identically priced Japanese models are available (p 17). For larger whole-house and whole-building systems, detailed Danish studies found that the best marketed 1988 equipment could save 45 per cent of the energy typically used in 1988 for this purpose, while more advanced equipment could raise the saving to 85 per cent. However, even this proved conservative: the best Singapore practice in very large buildings, which should be more efficient to start with, saves typically around 90 per cent and reduces capital costs (Chapter 1.17, p 53).
- The Danish analysts assumed only a 30 per cent short-term and 50 per cent advanced-technology saving from other appliances, the most important being televisions. However, this is clearly conservative. Much larger savings are commonplace in today's market, with no correlation between efficiency and equipment price.

What happens when all these available and cost-effective savings are added up? Electricity use is reduced by 74 per cent with advanced technologies already clearly feasible in 1988 and generally on the market (if not surpassed by the best on the market) by 1994. These measures alone would maintain or improve service quality while decreasing Danish appliance and lighting usage per capita from the 3,200 kWh/y average to only 825 kWh/y. Adding some cost-effective substitution of heat sources other than electricity would raise these savings to 80 per cent – a value also readily achievable for lighting (Chapter 1.11, p36) – and reduce per capita usage to only 620 kWh/y. And as the Danish experts emphasise, the cost of such electrical savings, per kWh saved, is less than 'that of marginal production of electricity in any place in the world.'

A separate analysis by the same team in 1983 found cost-effective potential savings of 72 per cent in Danish cooling electricity and 65 per cent in Danish heating electricity. Both figures could be considerably increased today, thanks partly to superwindows (see p 19). Indeed, by 1987 an official Ministry of Energy analysis found a 66 per cent saving potential in Danish cooling and 62 per cent in ventilation, even assuming only technologies likely to enter the market spontaneously rather than speeded up by policy actions.

1.10 Super-refrigerators

Cold beer, fresh fish, crisp vegetables, sweet milk – these benefits of household refrigeration are familiar throughout the world's wealthier countries. Even more important, in the villages of the South, a small solar-powered medical refrigerator may mean the difference between life-saving vaccines and fatal diseases. But for either use, what does it really take to keep a modest volume reliably tens of degrees cooler than its surroundings? Just an insulated box and a cooling method – traditionally a device that alternately compresses CFCs and re-expands them, moving heat from the food to a 'condenser' coil on the outside of the box.

The main failing in most refrigerators is their insulation. From about 1950 to 1975, as electricity became cheaper, refrigerator-makers kept making the insulation thinner so they could make the inside of the refrigerator bigger without making the outside bigger. (Given time, they might have made the inside bigger than the outside.) They also used very inefficient compressors, often mounted underneath so the heat would rise into the food compartment and have to be removed all over again. The compressor and condenser were so badly designed, and the condenser so undersized, that they needed a noisy, inefficient fan blowing on them to prevent overheating. The condenser coil was complex and hard to clean, so dust would build up and keep the heat from escaping, making the compressor run more. The door wasn't well sealed against air leaks, and when it was opened to expose the open shelves, all the cold air would fall out. The thin insulation caused the outer surface of the box to 'sweat' with condensation in humid weather, so the manufacturers installed electric heaters to dry it out; until recent years these couldn't even be turned off in dry sites or seasons. The antisweat heaters teamed with the thin insulation to help the extra heat get back inside faster. Inefficient fans were added to distribute the cooling, substituting for good design in the first place, and to make sure the food got dried out as well as cooled. To reduce frosting inside, electric heaters were put there too, along with inefficient lights, just to make sure the cooling system had plenty to do. These heaters would run regularly even if there was no frost. Increasingly, too, options were added that fed ice or various drinks out through the door without having to open it. This slight convenience added another hole in the insulation.

What a ridiculous way to cool food! Hundreds of millions of units were nonetheless sold, and many are still in service. Each one wastes so much electricity that the coal burned to generate it – coal you pay to have needlessly turned into global warming and acid rain – would about fill up the inside of the refrigerator every year.

American refrigerators and combination refrigerator-freezers (the more common type) together used a sixth of all residential electricity in the late 1980s, equivalent to the output of about 30 Chernobyl-sized power stations. But meanwhile the new units, which replace old ones about every 15 years – units may last 20 years but are often resold before that – were becoming markedly more efficient.

Since the mid-1970s, and especially in the 1990s, manufacturers discovered how easy it is to use more insulation, better seals, better designs, bigger coils and more efficient lights, compressors and controls to keep the food just as cool and dry it out less while using far less electricity. The effect was dramatic. Per litre of volume (adjusted for the mixture of refrigerator and freezer space):

- The average model sold in the United States in 1972 used 3.36 kWh/y.
- By 1987, when efficiency standards came into force in California, this had fallen to 1.87 kWh/y.
- In 1990, a new Federal standard prohibited the sale of models worse than 1.52, and the best mass-produced model used only 1.32, yet cost less than the average new unit in its size class.
- In 1993, the Federal standard was tightened to 1.16.
- In 1994, Whirlpool won the Golden Carrot competition (an approach adapted from an earlier Swedish design contest won by Electrolux) with a unit that used 1.08, and the major US manufacturers agreed to cut that to no more than 0.86 by 1998.
- Ever since 1988, Gram in Denmark has been producing better-insulated refrigerators using only 0.45, readily reducible to about 0.26 by further improvements.

Some specially made equipment for niche markets did even better:

- Even in the early 1980s, a small American firm, Sun Frost, was hand-making (mainly for solar-powered households eager to minimise their purchases of costly solar cells) models using only 0.45–0.53 – units that cost more because of low manufacturing volume, but that in mass production should cost even less than conventional models because they're so much simpler. (Current versions with partial defrost, replacing the older manual-defrost models, use 0.60–0.70.)
- Since 1983, a Sun Frost refrigerator at Rocky Mountain Institute has used only about 0.19, because half its cooling is passively provided by a 'heat pipe' connecting it to a shaded outdoor cooling fin, releasing the food's heat into the often cold outside air.

One might suppose that the 86 per cent reduction in energy per litre from the 1972 US norm to the early-1980s Sun Frost would be about as far as was feasible and worthwhile. Far from it. Many further options beckon:

- At least five kinds of more advanced insulating material can insulate 2–12 times as well per unit thickness as the best plastic foam, which in turn is twice as insulating as glass fibre/mineral wool. Perhaps the most intriguing of these new materials is simply two sheets of stainless steel, edge-welded a few millimetres apart (separated by little glass balls), with a hard vacuum inside and with the inside coated with a special film to block heat-robbing infrared rays. A cardboard-thin layer of such 'compact vacuum' insulation can stop heat flow as well as 7 cm of mineral wool. It costs more, but it can also superinsulate the refrigerator while making its walls much thinner. The increased interior volume made possible by the reduced thickness is worth about enough to pay for the exotic insulation.
- Compressors are normally too big and inefficient: even Sun Frost's aren't the best, because the company couldn't afford bulk purchases of more efficient models. But Sun Frost does use separately optimised compressors for the refrigerator and freezer, and mounts them on top. Many further improvements are available, including variable-speed compressors that cool at just the rate required.

- New Stirling-engine compressor designs can boost efficiency by one-half to two-thirds. They scale down well to the very small sizes appropriate for superinsulated refrigerators. They also improve reliability, reduce noise and eliminate chlorine-containing refrigerants in favour of inert helium.
- The condenser coil can get bigger by bonding it to a heavy metal plate in the unit's cabinet, nearly doubling the refrigerator's efficiency. Gram's engineers developed this concept initially because adding weight in the back would keep the unit from tipping over forward if a child hung onto the door.
- Better materials and designs can reduce air leakage through seals.
- Better fans and lights can cut operating energy and heat food less. Lights can even feed only light inside – not heat – through light pipes or fibre optics. Some models, like Sun Frost and Gram, design better to eliminate interior fans, especially in the refrigerator compartment, helping keep food fresher. And well-engineered units use a big enough condenser coil, mounted on top, to eliminate its fan as well.
- New sensors can detect frost and defrost only when needed, reducing defrost energy about tenfold.
- Defrosting can be done by fluid warmed by wasted condenser heat, rather than by electricity.
- All the condenser heat can be used to heat water for the household.

It appears that using these improvements could increase the already-achieved 86 per cent savings to at least 95 per cent – with no loss of service, reliability or (probably) money. The super-refrigerator revolution has only just begun.

1.11 LIGHTING

One-fifth of all electricity used in the US goes directly into lighting – actually, one-quarter when we add the energy used to take away the heat of the lights. That is about as much electricity as more than 120 giant power plants make. In countries like Russia or China, roughly 15 1,000 MW power stations are fully occupied just running inefficient lights.

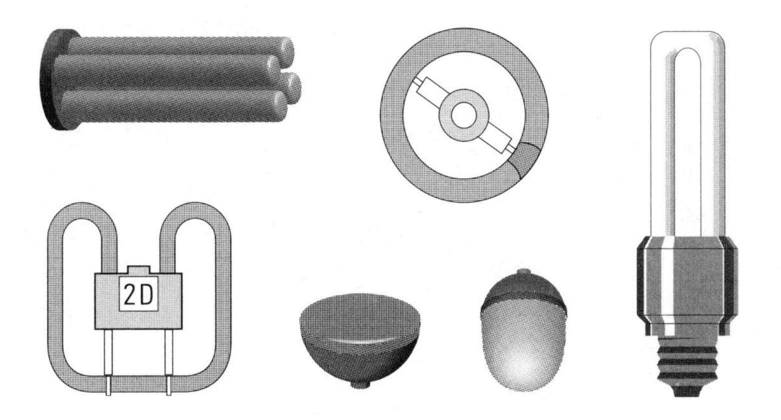

Figure 4 Compact fluorescent lamps can be obtained in a multitude of shapes – over 500 in the US alone.

Incandescent Lamps

Close to half the US lighting energy, and a higher fraction in most developing and formerly socialist countries, is used by ordinary incandescent light-bulbs, which have advanced only modestly since the 1930s. Such lamps are actually electric heaters that happen to emit 10 per cent of their energy as light. Nearly all of them can be directly replaced by compact fluorescent lamps – pioneered in Holland and Germany, introduced to the North American market in 1981, lately brought into local production in Eastern Europe and China, and now selling over 200 million units a year worldwide, rising by 15–20 per cent per year. The ones sold in 1994 alone will save at least $5,000 million worth of electricity over their lifetimes.

That may not seem like many lamps in a world that consumes nearly 10 thousand million incandescent lamps per year. But the compact fluorescent lamps also last about ten times as long, so their 200-million-unit-per-year sales are equivalent to about 2 thousand million incandescent lamps, or about a one-fifth global market share in terms of the amount of light delivered. This difference in lifetime also means that once half the sockets in the world hold compact fluorescent lamps, they would still amount to only about 5 per cent of the bulbs being sold. But at least in business uses, where people count the cost of replacement lamps and the labour needed to install them (usually in the ceiling), the increased lamp life more than pays for the compact fluorescents themselves. That makes their electricity savings better than free: not a free lunch, but a lunch you're paid to eat. Even counting just the replacement lamps, not their installation labour, the compact fluorescent lamp repays its cost many times over during its lifetime.

Compact fluorescent lamps also illustrate how pollution is prevented not at a cost but at a profit when it's cheaper to save energy than to produce it. A single 18-watt compact fluorescent lamp, replacing a standard 75-watt incandescent lamp, can save over its lifetime (Lovins, 1990):

- a tonne of carbon dioxide, 4 kg of sulphur oxides, and 1 kg of nitrogen oxides, among other emissions from a coal-fired plant; or
- a half-curie of strontium–90 and caesium–137 (among other high-level wastes), and 0.4 tonne TNT-equivalent of explosive plutonium, from a nuclear plant; or
- at least 200 litres of oil fed into an oil-fired power station (*e.g.*, in many developing countries) – enough to drive an ordinary family car 1,000 miles (1,600 km), or a hypercar from one to five times across the US.

To put it another way, the fuel energy saved by replacing eight incandescent lamps, operating continuously, with such compact fluorescent lamps is enough to fuel typical driving in an average American car. Moreover, a $7.5-million factory produces up to 5,000 compact fluorescent lamps per day. The electricity saved by the lamps produced by that plant also avoids building power plants costing at least 40 times more (hundreds of times more in a country like India); or saves as much energy as comes from an offshore oil platform costing several hundred million dollars; or saves as much energy as is used by 188,000 American cars, or by six fully laden, fuel-efficient Boeing 757 passenger jets engaged in continuous long-distance service.

Such compact fluorescent lamps can also, for example, cut the peak load in Bombay by a third, stretching meagre power supplies; or raise a North Carolina chicken farmer's profits by one-quarter; or increase disposable income in a very poor country like Haiti by perhaps as much as one-fifth. Not bad for a little device you can hold in your hand and screw into the socket yourself.

Compact fluorescent lamps are not the only option. Large incandescent lamps are often best replaced with metal-halide or high-pressure sodium lamps, some of which now produce pure white light virtually indistinguishable from daylight. Where a concentrated beam of light is required, as in retail displays, options include lamps carefully designed to reflect light from a small quartz-halogen capsule – while using thin films like those in superwindows to reflect heat back onto the filament so it needs less electricity to remain white-hot. This design enables 60 watts to produce exactly the same light that used to require 150.

Tubular Fluorescent Lamps

Half of US lighting energy, or about $360 per person per year – and an even higher fraction in many other Western countries – is used by tubular fluorescent lamps and the 'ballasts' that start and control them. But around 80–90 per cent of that energy is wasted, and the light produced is quite unsatisfactory. The main improvements normally needed are:

- ❑ Let a lot more of the light escape from the fixture into the room. Specially shaped, shiny materials inside the fixture can nearly double efficiency in most cases, while better controlling how much light goes in which directions to improve visibility and reduce glare.
- ❑ Design or modify the fixture to keep the lamps and ballasts at the best temperature: they're usually too hot, making them lose more energy and not last as long.
- ❑ Use lamps that emit just the right colours to match the eye's red, green and blue receptors. This makes colours more accurate and attractive, and actually helps the eye see better with less light.
- ❑ Use more slender lamps that emit up to one-quarter more light per watt and make it easier to design optics to control where the light is going.
- ❑ Operate the lamps at high frequency, eliminating the flicker and hum that for many people cause fatigue and headaches. The electronic ballasts that run at high frequency also save at least one-quarter of the energy directly, and considerably more when more subtle effects are counted.
- ❑ Use electronic controls to dim the lamps according to how much daylight is present, turn them off when there's already enough daylight or nobody is in the room, set them at just the brightness that's right for that part of the room and that people want for the tasks they're doing at the time, and automatically brighten the lamps as they dim with age or dirt (so

it's no longer necessary to have too much light when the lamps are young, fresh and clean in order to be sure of having enough when they're old, dim and dirty).
- Keep the lamps and fixtures properly cleaned, and replace lamps regularly before they grow dim or fail.

LIGHTING DESIGN

Together, these measures will at least quadruple the efficiency of typical fluorescent lighting systems and will pay for themselves in a few years. But other very large and even cheaper savings are often achievable by improving how the light is used:

- Make the visual task easier by such simple measures as fixing that blurry photocopier.
- Arrange the room so the light doesn't glare straight off the page into your eye: what lets you see isn't light but contrast (*e.g.*, between ink and paper), and 'veiling glare' washes out the contrast. In a typical office, it's about ten times more important to reduce veiling glare than to add more light.
- Consider bouncing the light indirectly off the ceiling or wall, so it comes from many different directions and thus virtually eliminates veiling glare. Such 'indirect' lighting can let you see better with only one-fifth as much light.
- Provide the right amount of light for the task – not too much as is commonly done. The right amount depends on an individual's eyes and age, the time of day, the difficulty and importance of the task, and other factors. This makes it important for people to be able to adjust lighting levels to their current needs.
- Instead of overlighting the whole space, use less general or 'ambient' lighting and add 'task lighting' just when and where you need it. For example, a task light can help you read papers next to a computer screen without over-lighting and washing out the screen so it becomes unreadable, and without having such a high ratio of brightness between the screen and the paper copy that your eyes get tired switching back and forth.
- Use lighter-coloured carpets, paint and furniture to help light bounce around the room better.
- Bounce daylight deeper into the building, by methods ranging from top-silvered venetian blinds to special 'lightshelves'

(which can now move light as far sideways as desired – even tens of metres), and deliver the daylight evenly and without glare. Direct solar rays are often so bright that they make it harder and more tiring to see; instead, daylight should generally be sent upwards so it lights the ceiling. Glass-topped partitions can keep individual offices private, yet spread daylight everywhere.

- In difficult cases, concentrate sunlight outside, perhaps on the roof, then bring it into the core of the building with atria, clerestories, light shafts, light pipes, fibre optics and other modern methods. (By such means, Japanese architects can even bring bright daylight many storeys underground.)

Combined with better lighting equipment, these and other lighting design techniques can usually save more than 90 per cent of lighting energy very cost-effectively, yet also look nicer and enable people to see much better. This in turn can dramatically improve how much and how well people work.

1.12 Office Equipment

In most of the industrial world, the fastest-growing use of electricity, in the fastest-growing sector (commercial buildings), is office equipment. This is a natural result of the rise of the information economy. A substantial fraction of office equipment, too, isn't in offices, but at retail checkout counters, in hospitals and schools, and everywhere else people need information.

Computers

An inefficient modern desktop computer and screen use electricity at a rate of 150 watts whenever it's turned on. (What the computer is doing makes almost no difference.) Typically at least half of that use is in the colour monitor, which is rather like a colour television. But careful shopping for colour televisions discloses that fourfold less electricity, or even less, is used by the most efficient models with the same size, features and price as the least efficient ones. Is the same true for computers too?

Certainly, and for the same reason: the quality of design. Some kinds of computer chips and power supplies use far more energy than others. Hard-disk drives even 5 years old can easily use 5–10

Figure 5 A modern laptop computer can ideally reduce electricity demand by 90–99 per cent when compared with an old-fashioned desktop computer with similar capabilities.

times as much energy as modern ones that work better and cost less. Portable computers, designed to eke out long life from lightweight batteries, use only a few watts, yet can be as powerful as those 150-watt desktop computers: for example, this is being written on a subnotebook computer that uses only 1.5 watts, or 1 per cent of the norm for an inefficient desktop machine with exactly the same capabilities and much less portability. It runs for 6–9 hours on just 5 oz (150 g) of nickel-metal-hydride batteries, or 100 g of lithium batteries. Some full-fledged 'palmtop' computers even run for a month on two little AA alkaline batteries!

Part of the difference is in how the energy is managed. Canadian researchers found that computers are not actually being used, as measured by keystrokes, for about 90 per cent of the time they're turned on. Devices and software can be added to most existing computers to put them into a sort of sleep or hibernation until they're needed again – when they spring instantly back to life at the touch of a key. Portable computers simply build in that power-management feature, turning off various parts whenever they're not needed. Some models even slow the main processing chip to a crawl, or suspend its activity, whenever it's not needed – if only for a period as short as the instant between keystrokes.

Such efficient components and power management don't make portable computers cost more (except for flat-panel colour displays); in fact, some portables now cost the same as their desktop equivalents, or even less because they save materials. Most manufacturers make both kinds, so to simplify their production, they're now starting to use the same efficient components and system designs in both kinds; all that really differs is the box, the room in the box to add expansion options and the type of display. Moreover, portable computers add to their convenience – for taking your work on the train or down the corridor – a hidden economic advantage: operating always on a built-in battery, they don't require a special 'uninterruptible power supply' and special wiring to deliver its glitch-free power to each desk – an avoided investment often totalling hundreds of dollars per worker.

Design Synergies

This became obvious a few years ago when a major computer manufacturer wanted to build a desktop computer with the same features – very compact box, flat-panel colour display, lots of slots for credit-card-sized accessories – as a notebook-sized computer, and with comparable energy efficiency. The first task was to improve the power supply. Nearly all are made by a few Asian firms with similarly poor designs: their efficiencies are often below 50–60 per cent at high loads, and fall off disastrously at low loads; but most of the time they're at low loads because the user hasn't installed all the power-hungry accessories they're sized to serve.

It turned out that for a slightly higher price, an efficiency in the mid-nineties of per cent could be achieved, across the full range of loads, along with much higher power quality, which utilities like. The bean counters objected to the higher price. But then the designers realised they could save more money than that by eliminating the fan: the power supply and the chips and drives it powered were now so efficient that they could cool by natural convection. Moreover, the power supply shrank to such an unheard-of small size that the whole box got smaller too, saving materials and cost. Then the marketers realised they'd stumbled onto a bonanza: they could sell the computer as the first desktop model to run silently* because it had no fan; to take up very little desk space; to be more reliable, because, without a fan, there was no air path through the

* Well, almost silently: now for the first time one can hear the annoying mosquito-like whine of the hard drive when it spins up. But this sound, previously masked by far louder fan noise, can in turn be eliminated by using the existing chips and loudspeaker to cancel it out with 'antinoise' at virtually zero extra cost: Japanese refrigerators without those free components already add this feature at a cost of only a few dollars.

machine to deposit dust on the chips and eventually cause them to overheat; and even to offer users the option of locking the valuable but very small computer in a desk drawer for security.

ENERGY-EFFICIENT IMAGES

This is not the only kind of office equipment that can save nearly all its energy while costing no more (or even less). Printers, fax machines and other 'imaging' equipment typically use even more of an office's electricity than do computers and monitors. Most imaging in a modern office is done with devices that use a laser to form an image on a photosensitive drum, then use the standard xerographic photocopying process that ends by melting plastic toner powder onto the paper with a hot drum. Heating the drum takes many hundreds of watts, heating the office whether it's needed or not. The laser printer is also a very precise electro-optical device, often involving a spinning mirror, precise lenses and other complexities.

But modern ink-jet printers don't need that hot drum. Instead, they use microscopic currents flowing in a walnut-sized print head to boil infinitesimal amounts of fast-drying ink, spraying minute droplets onto the paper to form the image. The sophistication is in the print head; the printer mechanism itself is very cheap and simple, needing only to move the paper, and even the print head doesn't cost much, because it's mass-produced like chips (and, often, refillable with fresh ink). Ink-jet printers and fax machines use only 1 or 2 per cent as much electricity as their laser equivalents, yet they produce about the same image quality with a speed, per typical print job, that's usually comparable. They're also smaller, lighter, more reliable and only about half as dear.

Or consider photocopiers – the biggest electricity hog in a typical office. At Rocky Mountain Institute a few years ago we saved a third of the energy in a standard new photocopier just by shopping carefully for better engineering. It also cost 15 per cent less. More recently, we saved more than half of the remaining energy, at still lower capital cost and better reliability, by switching to a newer model that has no standby energy, because its fuser (which normally melts thermoplastic toner powder onto the paper) is not a metal roller but rather a rubber belt that doesn't start to heat up until the paper is actually approaching it. We also wanted a smaller copier that people could just walk up to and use to make instant copies without any warm-up delay, so we got a second-hand older model that squeezes waxy toner powder onto the paper with a cold compression roller using no heat at all. It saved 90 per cent in both energy and capital cost, and is far more reliable than the hot-fuser models. Large, high-speed versions are already widely used to print mass mailings like credit-card bills.

That's just the beginning of what's becoming available. New kinds of toner should soon be able to fuse with a flash of ultraviolet light rather than having to be melted onto the paper. And many manufacturers have recently introduced machines that make many copies of a given document with a non-xerographic process – more like the old duplicating machines, but attractive and all-digital – using only 1 per cent as much energy as a photocopier.

Power management is also being rapidly added to imaging and copying equipment. Almost all new laser printers and computers now comply with the US Environmental Protection Agency's voluntary Energy Star standard, which requires power-saving standby modes. (President Clinton ordered federal agencies to buy nothing else without a special reason, and many private companies adopted similar policies.) Now that almost all manufacturers comply, the next step is to bring the standards more up-to-date to reflect today's far better technologies. Other voluntary efforts are making it easier; for example, for computers that are 'asleep' to wake up to take incoming modem calls or backup requirements, rather than having to be left on all night in case they're needed.

The Benefits Build Up

What do these savings add up to? Retrofitting power management onto existing equipment, and training people to turn off anything they won't use for a while, can usually save at least two-thirds of the energy. Buying the most efficient new equipment saves more like 80–90 per cent, and with careful shopping can save closer to 96 per cent, using equipment that works the same or better and costs the same or less. In the US alone, that can save the equivalent of dozens of giant power plants over the next few decades.

More efficient office equipment, multiplied by its millions of units, can also save building owners enormous amounts of construction cost for wiring, cooling and ventilation. Specifying very efficient equipment in a typical large new office building can cut that building's total construction cost by as much as 6–8 per cent – enough to justify buying all-new office equipment for that purpose, even though normally existing equipment might not be replaced for another few years. Efficient office equipment, like efficient lighting, can also help avoid the painfully high cost of increasing wiring or cooling capacity in older buildings not designed for all that modern office equipment. All told, a single energy-efficient desktop computer can save society costs totalling from one to several thousand dollars (Lovins, 1993) – about as much as the entire computer costs!

1.13 PHOTOVOLTAICS AT FORTY-EIGHT VOLTS DIRECT CURRENT: EDISON'S GENIUS REDISCOVERED

Thomas Alva Edison (1847–1931) was the greatest technical inventor of his time. He invented the (carbon filament) light-bulb, the microphone, a greatly improved telephone, an early version of the gramophone and the first movie camera. And Edison linked up the steam engine with large electricity generators which served as the basis for the first public electric utility, founded in New York in 1882 – by Edison himself.

A great annoyance to Edison after he had invented the power station was the progressive displacement of low-risk and efficient-to-use low-voltage direct current (DC) by high-voltage alternating current (AC), either at often lethal 220–230 volts or at 110 volts that is still harmful and often less efficient. (At the time, he didn't even know about the electromagnetic smog that is a by-product of AC alternating current.) The victory of AC resulted from the effort to reduce transport losses in electricity mains. For efficient power transfer over large distances in cables with limited cross-sections, very high voltages are required, *e.g.* 50,000 volts. For end-use, this high voltage has to be transformed back down to low voltages, *e.g.* 110 or 220 volts – something physics won't permit with DC.

There are two reasons why AC is more wasteful in many appliances. First, reversing the magnetization in electric motors some 100–120 times per second, more than in a classical DC motor, just generates a lot of heat in the iron. Second, transforming AC into DC is a wasteful process; just try touching the hot external transformers of any household electronics equipment.

A 20-watt AC water pump can be replaced by an 8-watt DC water pump of equal performance in a two-and-a-half-fold lower consumption of raw electric power. For computers, video recorders or fans, potential savings are even more impressive: here the use of DC could be 6 to 10 times more efficent than that of AC. For household appliances such as refrigerators and television sets, the improvement in efficiency gained by using DC (*i.e.* without the improvements mentioned in Chapter 1.9) would still be some 60 per cent.

Friedrich Lapp, Günter Scharf and Gerd Ehrmann of the Nürnberg Vocational School No. 1 decided it was high time to exploit the advantages of DC and benefit from the wisdom of the great Edison, although their specific motive could not have been imagined by Edison: they were excited about photovoltaics. Solar cells, however, are expensive. In order to generate in cloudy Germany the electricity needed by a typical family of four using normally inefficient 220-volt AC appliances, at least 30 sq m of solar cells are required, costing some $50,000. Using DC appliances, some 8 sq m of solar cells costing $15,000, plus a passive solar water-heating system for an extra couple of thousand dollars, would be sufficient. Efficient appliances further reduce the needed pv area.

The Nürnberg solar team, whose laboratory by now is a well-known attraction for vocational students at the school, then went a step further and looked into the optimal voltage for DC supply.

At 12 volts, the voltage of car batteries, big copper wires (24 sq mm cross-section) would be necessary to cope with the power requirements of an ordinary household, but cables that thick would be costly and would carry with them a heavy ecological 'rucksack' (see p 242). At 24 volts, the necessary cross-section would be reduced to 6 sq mm and at 48 volts to an agreeable 1.5 sq mm.

Thus, the Nürnberg team seems to have identified an exciting break-through strategy for photovoltaics and for exploiting the efficiency potential (and its safety advantages for families with small children) of low-voltage DC in private homes. In terms of CO_2 abatement, their strategy might yield far more than a Factor Four. The trouble is, however, that almost no industrial production exists of 48-volt DC appliances – only 12 (rarely 24)–224-volt for boats, caravans, etc. Manufacturers argue that there is no demand. Small wonder in countries where 220-volt (or 110-volt) AC is conveniently available in every room and where the winter low point in photovoltaic electricity generation makes add-on electricity from the power grid desirable.

The breakthrough for Edison's wisdom in our times might take place in countries without an extended power grid but with reliable sunshine all year round, or where small-scale wind- and hydropower is available. In those countries, the DC efficiency path seems far more rational than the construction of a wasteful AC countrywide infrastructure.

But let's be honest. If we in Europe or North America were poor, and the rich people of this world were to demonstrate to us the life they led with a centralized power supply day by day and swamp us with offers to install the same supply system in our country, surely we would succumb and imitate the wasteful ways of the rich.

1.14 RENEWABLES IN A COLD CLIMATE

Niels Meyer *et al.* (1993) estimate that for the Scandinavian countries a 95 per cent reduction of CO_2 will be necessary as a fair contribution of those countries to stabilising the earth's climate. Sustainable energy development, according to these authors, would comprise four strategies:

- improved technologies (meaning the efficiency revolution),
- environmentally benign energy sources (meaning renewables),
- structural changes, especially in the transport sector, and
- reduced growth of energy services.

The authors are more or less in line with us regarding the potential of the efficiency revolution. They propose reductions of total primary energy consumption for Denmark by 79 per cent (more than a factor of four), for Norway by 59 per cent and for Sweden by 54 per cent. For the three countries taken together, the reduction is 66 per cent. Table 1 shows the scenario results for Norway.

The scenario is remarkable in many respects:

- Direct solar energy (photovoltaic and passive solar) is not assumed to play a role, because in Norway practically all space heating and hot-water use in the household and service sectors are fuelled with electricity, which is cheaply available. In virtually all other countries direct solar energy would be a non negligible quantity.
- Total energy consumption is not assumed to be reduced by a factor of four as is assumed for Denmark. The reason is that Norway is in an exceptionally comfortable position regarding hydropower and can thus be assumed to maintain its position as a good place for highly energy-intensive industries such as aluminium smelting.
- Nevertheless, more than 60 per cent of Norwegian hydropower is assumed to be exported to other countries.
- The private car fleets of the Scandinavian countries are assumed in the scenario to consist mostly of high-efficiency electric, hybrid or fuel-cell cars.
- Finally, the scenario (for all Scandinavian countries) assumes that the levels of energy services will not increase until 2030. This is the result of the uncompromising postulate of achieving until 2030 a level of globally sustainable per capita energy

Table 1 Primary energy consumption in TW (thousand million kWh) per year for Norway in 1987 and calculated from the scenario model for the year 2030 (rounded numbers). (From Meyer et al, 1993, p 134)

Units of TWh/y	*1987*	*2030*	*2030 as % of 1987*
Biomass	11	9	82
Hydropower	104	112	108
Windpower	0	12	∞
Wavepower	0	10	∞
Direct solar energy	0	0	–
Natural gas	0	3	∞
Oil	26	0	0
Petrol	22	0.3	1
Diesel	32	1.2	4
Coal	0	0	–
Electricity import	3	0	0
Electricity export	3	68	2,267
Total consumption	195	80	41

consumption. In the real world, it will be extremely difficult to fulfil this ambitious postulate.

Although the total reduction in energy consumption in the scenario is less than a factor of four, we felt we should include it in our book, for the following reasons:

- Renewable sources of energy are in a sense equivalent to efficiency gains. Under the criterion of 'carbon efficiency', the scenario represents roughly a 30-fold improvement.
- The study offers perhaps the only well-calculated scenario for all sectors of a modern economy taken together, and that not only for one country but for a number of countries with quite differing geographical and demographic conditions (Denmark has no water power at all and is densely populated).
- If Scandinavia in total is able under the scenario conditions to export nearly 30 per cent of its energy supplies to other countries, it would help those to reach sustainable levels of CO_2 emissions.
- 'Factor Four' has not been the explicit goal of the study. As

the Danish example seems to indicate, it would have been quite possible to reach that target for each country.

Although the study by Meyer and his colleagues from Sweden and Norway is extremely valuable and encouraging, it is not directly transferable to other countries. The Scandinavian situation is special regarding hydropower and low population density. Generally, renewable sources of energy should not be seen as a cure-all.

1.15 LOW-ENERGY BEEF

Farming has been the traditional source of energy harvesting for humans. In traditional societies some 80 per cent of the total human-induced energy flow was calories contained in food. Although peasants and farm animals put in some mechanical energy, the input/output ratio was close to 1:100. Everything changed with the advent in the 20th century of agricultural mechanisation. Starting in the US, farming increasingly became a net energy consumer. In modern rice and wheat production, the input/output relation is between 0.1 and 0.4 (1 calorie input for 10 or 2.5 calories output, respectively). In fruit and vegetables it is between 0.5 and 10 (10 means 10 calories of input for 1 calorie in the food). But it may reach extreme values of 500 for winter greenhouse vegetables that are a common product from the Netherlands (see p 51).

Generally speaking, the ratio is better for plant products than for animal products. In milk, it ranges from 0.8 to 8, in eggs from 0.5 to 10, and in meat from 0.5 (for free-range chickens living mostly on what they find on the farm) up to 35 (for industrialised beef production using animal feed from overseas). Even fishing is a net energy consumer, which is surprising since the growing of fish requires no human energy inputs. The ratio ranges from 1 (extensive coastal fishing) to 250 for highly mechanised ocean fishing (see Plate 5). All figures are based on Immo Lünzer's (1992) classic study.

For increasing energy efficiency in agriculture, the strategic point of entry should be beef. It is a mass product and plays a pivotal role in modern agriculture. The simplest answer is to desubsidise and thereby reduce European overproduction of beef. Only through massive interventions including export subsidies for beef can farmers in Europe grow cattle that feed (mostly) on maize and overseas soyabeans, fishmeal, slaughterhouse waste and other rather unnatural fodder. Reducing export subsidies would save taxpayers huge amounts of money and radically reduce energy demand from the farms. Farmers would be induced to return to more ecologically acceptable methods of farming and to produce perhaps 50 per cent less beef in Europe. Consumers would pay higher prices per kilogramme but less money per month for smaller amounts of tastier and healthier beef.

The *Global 2000 Report* (Barney, 1980) presented a flow diagram of energy for American food production. For some 3.6 GJ (per capita) of human food energy, 35 GJ of technical energy have been expended, not counting the 'solar gift' of 80 GJ that is absorbed by the plants used in the process (Figure 6). It seems more than plausible that the energy 'demand' from agriculture and food processing can be reduced by a factor of four with essentially no sacrifice of wellbeing.

1.16 Greening the Red Ones*

The Netherlands have become one of the largest exporters of tomatoes in the world. This is not exactly what climatic conditions have assigned for Holland. The nightshade plant was imported to Europe from South America in the late 16th century for ornamental purposes. In Northern Europe it became a mass vegetable for food only during the 20th century.

The transition into a Holland-grown mass product came after the discovery in Holland and offshore of major natural gas reserves. Immense greenhouses were built, heated by natural gas, allowing tomatoes (and flowers and many other plants) to grow year round. In 1991, the Netherlands produced some 650,000 tonnes of tomatoes worth approximately £400 million from 1,600 hectares of greenhouses.

The huge quantities of tomatoes required a special marketing system. Today, Dutch tomato auctions attract producers from all around Europe, even including the Canary Islands. Roughly 15 per cent of the tomatoes marketed are consumed in the Netherlands; the rest are exported, including to Hungary where tomatoes are also grown under much more suitable climatic conditions than in Holland. But the Dutch produce tends to be cheaper.

* Acknowledgement is made to Wouter van Dieren and Geert Posma.

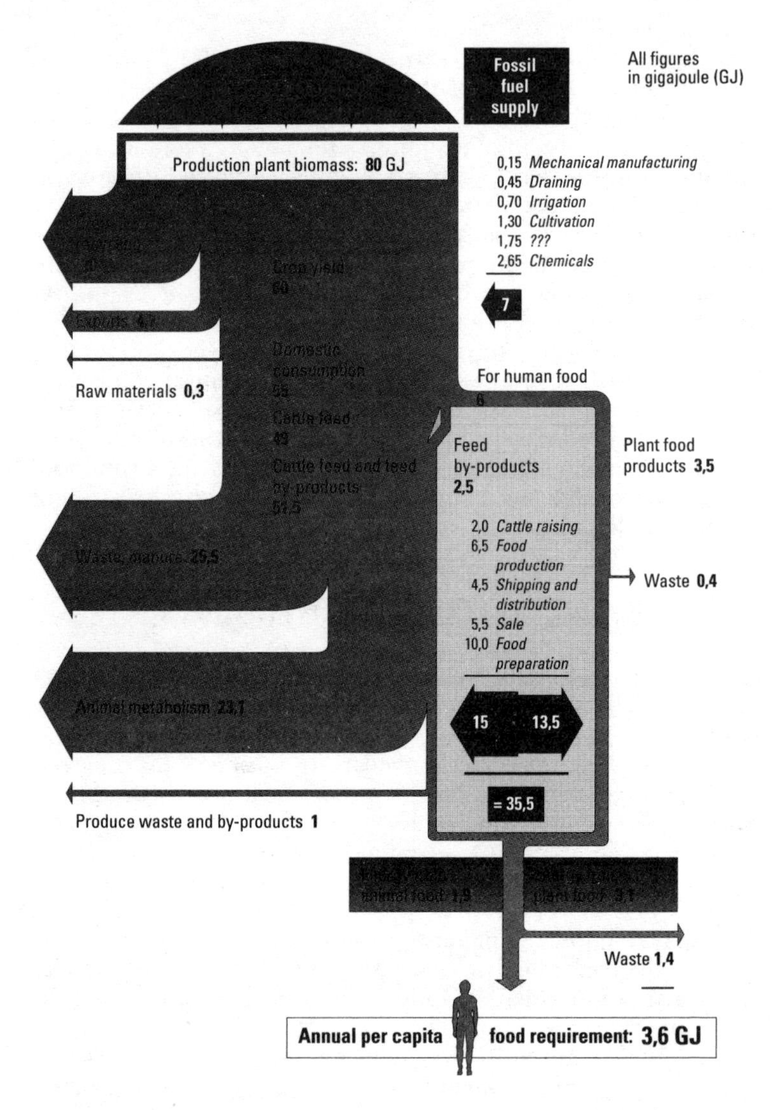

Figure 6 Energy in the food chain in the US. For producing an average 3.6 GJ of energy contained in food, some 35 GJ of energy are invested. (After Barney, 1980, p 295)

The reason for the marketing success, according to environmentalists, is the low price for energy. Only at that low energy price can tomatoes be grown at an energy input/output relation of 100 or more. Seventy-nine per cent of the energy used goes into heating the greenhouses; around 18 per cent goes into food processing, mostly in canning factories.

Is it possible to reduce this waste of energy? Surely the greenhouses could be much better insulated, without even adopting superwindow technology. According to Wouter van Dieren and Geert Posma, a factor of four should be attainable for Netherlands-grown tomatoes using existing technologies. More could be achieved if the fruit were grown (like bananas or mangoes) in countries that enjoy a more suitable climate. Even air-freight shipping of the tomatoes to Holland from, say, Sicily would cost less than one-third of the energy of the Dutch greenhouses.

1.17 Fans, Pumps and Motor Systems

In the industrial section of Singapore, a quiet, dryly humorous Chinese engineer* designs the world's most efficient air-conditioning systems (see Plate 6). Singapore has a difficult climate: relative humidity 84 per cent every day of the year, with temperatures ranging from hot to broiling. Most engineers would count themselves lucky to use only 1.75 kilowatts of electricity to provide 1 ton** of cooling there, and many use 2 or more. Lee Eng Lock's systems, however, use only 0.61 kW per ton – 65–70 per cent less. That's not an estimate. It is meticulously measured, once per minute, with hand-calibrated sensors that send six-significant-figure signals into his firm's extraordinary monitoring software.

Lee's systems also provide much better comfort, take up much less space, are more reliable and generally cost less to build. They cost less partly because every part is exactly the right size, not too big.

Elegant frugality is Lee's watchword. Energy, money, time, metal, every resource is used in just the right amount and place and manner. There is no wasted effort, motion or investment. Actual needs are measured, not guessed. Energy is used over and over until almost nothing is left. When he was once complimented on an especially

---

* Lee Eng Lock, Technical Director, Supersymmetry Services Pte Ltd, Block 73 Ayer Rajah Crescent #07/06-09, 0513 Singapore, tel 65 777-7755, fax 65 779-7608.

** As noted earlier, 1 'ton' of cooling means removing heat at the rate of 3.52 thermal kW.

ingenious way to do this – to use outgoing air from a big building to predry the incoming air, via a simple device with no moving parts – he replied, in characteristically telegraphic English, 'Like Chinese cooking. Use everything. Eat the feet.'

Most engineers would suppose that the place to save air-conditioning energy is in the 'chiller' that produces the cold water, because that's the biggest single user of energy in the cooling system. To be sure, Lee saves a third of its energy, chiefly by using heat exchangers three to ten times bigger (normal ones are grossly undersized) and making the chiller spin at just the right speed. But that's only a fifth of his total energy saving. Two-fifths is in the big 'supply fans' that blow chilled air around the building, and the other two-fifths is in the pumps and in the cooling-tower fans that dissipate the heat to the outdoors.

Lee's supply fans use not the normal good practice 0.60 kW/t but only 0.061 kW/t – 90 per cent less. His chilled-water pumps use not 0.16 but 0.018 kW/t – 89 per cent less. His condenser-water pumps, which move heat out of the chillers, use not 0.14 but 0.018 kW/t – 87 per cent less. His cooling towers use not 0.10 but 0.012 kW/t – 88 per cent less. Where do these roughly tenfold energy savings, with improved performance, come from?

From common sense, whole-system engineering, healthy scepticism about traditional practice and rigorous application of accepted engineering principles often ignored in that practice. Above all, from ruthlessly eliminating friction wherever it can be found.

ASKING THE FIVE WHYS

Taiichi Ohno, the pioneer of lean, continuous-flow, just-in-time production at Toyota, shared Henry Ford's obsession with eliminating waste and Frank Bunker Gilbreth's habit of digging into root causes. Ohno wrote, 'Underneath the "cause" of a problem, the real cause is hidden. In every case, we must dig up the real cause by asking why, why, why, why, why.' Joseph Romm quotes the example: 'Why did the machine stop? There was an overload, and the fuse blew. Why was there an overload? The bearing was not sufficiently lubricated. Why was it not lubricated sufficiently? The lubrication pump was not pumping sufficiently. Why was it not pumping sufficiently? The shaft of the pump was worn and rattling. Why was the shaft worn out? There was no strainer attached, and metal scrap got in.' (Romm, 1994, p 28)

Fans and pumps must move air or water against friction. Where does that friction come from? Lee relentlessly chases down the causes of the friction, asking the Five Whys:

- The pipe in the original design has too much friction because it's too long and has too many bends. That happened because the engineer laid out the equipment first, then connected it with pipes that had to run through all sorts of twists and turns to get from A to B. (The pipefitters didn't mind: they're paid by the hour.) Instead, let's lay out the pipes first, then the equipment.
- The pipe has so much friction because it's rough inside when it should be smooth. Choosing the right material and surface finish can cut friction 40-fold or more.
- The pipe is also too small. How easily the water flows varies as nearly the fifth power of diameter. If the pipe's diameter were increased by only 10 per cent, its friction would drop by 37 per cent; if 20 per cent, then by 59 per cent; if 50 per cent, then by 86 per cent. So fatter pipe can almost eliminate friction. It does cost a bit more, but the first designer balanced that extra cost only against the value of the saved energy, and using old prices at that. He forgot that by using fatter pipe we can make all the expensive parts – pump, motor, inverter, electricals – at least twofold smaller and hence cheaper. That's a better buy than too-slender pipe.
- The pipe has so many valves because there's so much friction that some parts of the piping system get less flow than they should, so valves were specified to add friction, so that places with excess flow will get less and will send more water to the starved parts. Why not just make all the pipes big enough that the water will go where it needs to – just as we make the wire in our house fat enough to carry the current where it's needed, rather than 'balancing' its flow with rheostats?
- The valves have so much friction because they're the wrong kind: nobody noticed. Their friction then makes the flow more unbalanced, requiring more valves. And so on.

The same happens with the supply fans:

- The diffusers that spread the air into the room are inefficient, causing noise and friction.
- They're connected by ducts that are laid out with sharp bends,

not gradual 'sweet' curves, and are too small in diameter, and are too long because they're in the wrong place.

- The coils are designed wrongly, so they don't cool or dehumidify well, and they have 20 times the air friction they should.
- The filters are too small because someone thought that made them cheaper; it's actually cheaper over time to make them much bigger, which makes them last enormously longer before they need replacing. Then their friction changes from big to tiny.
- To fight all that friction, the fan is so powerful that it makes too much noise, requiring a silencer that adds even more friction.

Of course, basic improvements like these are just the beginning of the design process. Lee starts with how much flow is really needed, then asks how short and smooth and sweet a pipe or duct can deliver that flow, then finds the fan or pump that has just the right size and characteristics to deliver that flow most efficiently, then scours the earth for the finest British fan or German pump to wring out the last bits of inefficiency, then works back upstream through the mechanical drivetrain, the motor, the inverter (which runs the fan at just the right speed for the moment, not more), the electricals. At each step he avoids compounding losses. The parts become smaller, simpler, cheaper. It's all really quite simple: like any other sort of genius, just an infinite capacity for taking pains.

In design as elsewhere, virtue is rewarded. When Lee is through making the air-conditioning system and all its parts severalfold more efficient, they no longer need to carry away so much of their own waste heat. (For example, all the energy that the fan imparts to the air to make it move makes the air hotter and has to be taken away again.) Thus not having to keep 'recycling' air-conditioning systems' inefficiencies over and over again not only saves energy, it also lets the costliest components become even smaller – rather like the way weight savings snowball in hypercars (p 4). The smaller and more efficient the cooling components get, the smaller and more efficient they can become.

Motors Make the World Go Round

It doesn't stop with the fans and pumps, either. They're driven by electric motors. Rocky Mountain Institute showed in 1989 how to combine 35 improvements in between the electric meter and the input shaft of the fan, pump or other device being driven. These motor-system improvements involve the choice, sizing, maintenance and lifetime of the motors; four kinds of control systems; improvements to the systems that feed electricity to the motor and those that transmit its torque to the machine it's driving; and how the whole system is managed and maintained. Together, the 35 improvements can typically save about half the energy fed into the motor, even with no downstream improvements (Howe *et al.*, 1993). These savings, too, pay for themselves in barely over a year, because if you pay for seven kinds of improvements, you get the other 28 as free by-products. The US utilities' think-tank agrees with those conclusions (Fickett *et al.*, 1990).

Lee is good with motors and electronic speed controls too, but he doesn't yet capture quite all of this potential; his practice emphasises space cooling, not motor systems. (He gets most of the savings – thereby reducing cooling loads even further and making his cooling systems even smaller – because most of the motors are in places where they don't add heat back into the building.)

But this is a much bigger issue than Lee's practice. Motors use more than half the electricity in the world. The 35 motor-system improvements alone, if fully used, can therefore save over one-quarter of the world's electricity – about 160 giant power stations in the US alone, about fourfold cheaper than just fuelling an existing coal-fired station, even if building it cost nothing.

What Was All That Cooling for Anyway?

Moreover, the cooling load itself – the number of tons required – can first be reduced severalfold by combining superwindows, efficient lights and office equipment, and better architecture. Most American offices are designed to need a ton of cooling load for every 250–400 sq ft of floor space. Yet a good retrofit normally achieves about 1,000, and a state-of-the-art new design gets close to 1,200 sq ft – around three to four times less. At the same time, people become more comfortable, and the whole building costs several per cent less to construct, because it needs so much less air-conditioning equipment at a whole-system cost (half of it for ducts and pipes) of around $3,000 per ton.

Lee Eng Lock's achievements are stunning. Yet they are not the end of the road. Let's have a further look at even more efficient kinds of cooling equipment than giant centrifugal chillers, and even more clever ways to keep people comfortable. That's our next topic.

1.18 THE FRONTIERS OF AIR-CONDITIONING

Transforming buildings from solar ovens into walk-in electric refrigerators uses about 16 per cent of the electricity in the US, which many consider the coldest country in the world in the summertime – indoors. Worse, air-conditioning causes about 43 per cent of the US peak load on hot summer afternoons, when it busies more than 200 giant (thousand-megawatt) power stations, each costing a couple of thousand million dollars or so.

In 1982 alone, residents and businesses just in the city of Houston, Texas, paid $3,310 million 'for cold air, more than the gross national product of 42 African nations.'* And the practice is spreading – most alarmingly to East Asia. There, demand for air-conditioning among teeming and newly prosperous populations (who are rapidly forgetting about traditional cooling methods and clothing) is adding some 25,000–50,000 MW of peak load per year – enough to require capital investments fatal to other Asian development goals.

It was not always so. For at least eight millennia, people have ingeniously designed their shelters to reject unwanted heat. From Turkey to Tunisia, Cyprus to Mali, Algeria to Zululand, sophisticated passive cooling systems achieved impressive levels of comfort not often approached in the same regions' 'modern' buildings today. Windowless 11th- and 12th-century Anasazi pueblos in the American Southwest, for example, kept temperatures that varied only one-quarter as much indoors as outdoors. On the north coast of Australia, traditional tropical house designs can maintain indoor temperatures as much as 32F° (19C°) cooler than outdoors. Classical Persian and Greek buildings did the same or better. An entire Roman city was passively air-conditioned. Arab goat-hair tents are marvels of passive cooling.

Today, science and technology can do even better. Space-cooling energy can be reduced not twofold, not tenfold, but close to 100-fold, just by systematically applying the best modern methods in six simple steps, each rich in diverse and exciting opportunities (Houghton *et al.*, 1993).

COOLNESS IS THE ABSENCE OF HEAT

The first step is to keep unwanted heat out of the building.

* B. Burrough, 'In Houston, the Ubiquitous Air Conditioner Makes Tolerable an Otherwise Muggy Life', *Wall Street Journal*, 21 September 1983, p. 31.

Superwindows let in cool, glare-free daylight while almost entirely excluding heat. Daylighting and more efficient, better-designed lighting cut that important cooling load at least tenfold (p 36). Efficient office equipment does the same for that load (p 41). (Lighting and office equipment together will then contribute about three times less heat than people do – a load that can't be reduced, except perhaps by reducing stress and frenetic activity.) Similar care is needed for internal devices ranging from refrigerators and vending machines to water coolers and coffee machines. Most if not all of these improvements can be retrofitted, paying for themselves typically in about three to eight years, and more or less immediately if the windows need replacing anyway (p 21).

In new construction, good architecture is vital too. Making the building the right shape and pointing it in the right direction can often save a third of its energy at no extra cost; in one ACT² (see p 15) office building in Antioch, California, it recently improved the building's total energy efficiency by 38 per cent while costing one-sixth less than it saved. Shading, massing and heat-reflecting surface finishes (like the whitewashed cities of the Mediterranean) can combine with landscaping and vegetative shading: one sizeable tree is equivalent to about a dozen room air-conditioners. And in any building, insulation and reduced air leaks usually help.

Expand the Comfort Envelope

People's sensation of comfort depends on how hard they're working, how much heat escapes through their clothing, the radiant temperature they experience from their surroundings and the temperature, humidity and movement of the air. Each of these variables offers important ways to expand the conditions in which people feel comfortable.

For example, ceiling fans can maintain comfort at temperatures 5C° (9F°) higher. Office chairs with mesh seats (like Herman Miller's Aeron model) ventilate the body, heating it approximately 10–15 per cent less than insulating upholstered chairs. Superwindows greatly reduce uncomfortable radiant heat hitting the body. (Ceiling coils carrying cold water can even provide 'radiant cooling'.) Just removing a necktie saves society about $50 worth of air-conditioning and power-supply equipment; over one-third of America's major corporations have already, for various reasons, relaxed their previously formal dress codes.

Together, such simple measures can reduce cooling needs by 20–30 per cent or more. In addition, the body doesn't respond instantly: it takes time to heat up 70 kg of water, and more time for the nervous system to report discomfort. The Alberta government harnessed this response lag by changing the control instructions in its big buildings so that the air-conditioning couldn't come on in the late afternoon; but by the time the buildings became hot, the workers had gone home. The chillers therefore ran for four to six times fewer hours, saving a lot of energy and money, but there were no complaints of discomfort.

PASSIVE COOLING

Unwanted heat that cannot be avoided or ignored should next be removed by the normal functioning of the building itself, without special equipment. Natural ventilation, coupling to cool earth beneath the building, or radiation to the sky, can do this in most US and many other climates, as it did for millennia worldwide. Even in mid-August in Miami, officially defined comfort can be maintained with no more than ceiling fans and a roof pond that stores up heat during the day, then radiates it away to the night sky.

Some very effective methods are almost passive. For example, Davis Energy Group in Davis, California, has developed WhiteCap, a shallow roof pond beneath squares of white foam insulation. During the day, the building's heat is transferred into the water. At night, a small pump sprays the water up into the air so that it cools, two-thirds by radiation and only one-third by water-using evaporation. The cool water then trickles back down through the cracks between the insulating panels and stays cool beneath them. The electricity used for pumping is only a few per cent of the cooling energy saved. The extra capital cost is zero, partly because the roof membrane lasts severalfold longer when protected by the overlying water from ozone, ultraviolet, temperature swings, shoes and other insults. Combining WhiteCap with skylights, so that both cooling and lighting are passive, could save more than 90 per cent of total energy in the many one- and two-storey flat-roofed buildings in the western US, with improved comfort and no extra construction cost.

Another example of passive methods: ice ponds that store winter coolness through the summer. They can be as simple as freezing slush kept cool under a layer of straw. The melt-water, at freezing temperature, is simply pumped through the building, using only a few per cent of the energy otherwise needed to chill it. In sites with spare land, this can be cost-effective even in climates with only a week or two of frost per winter.

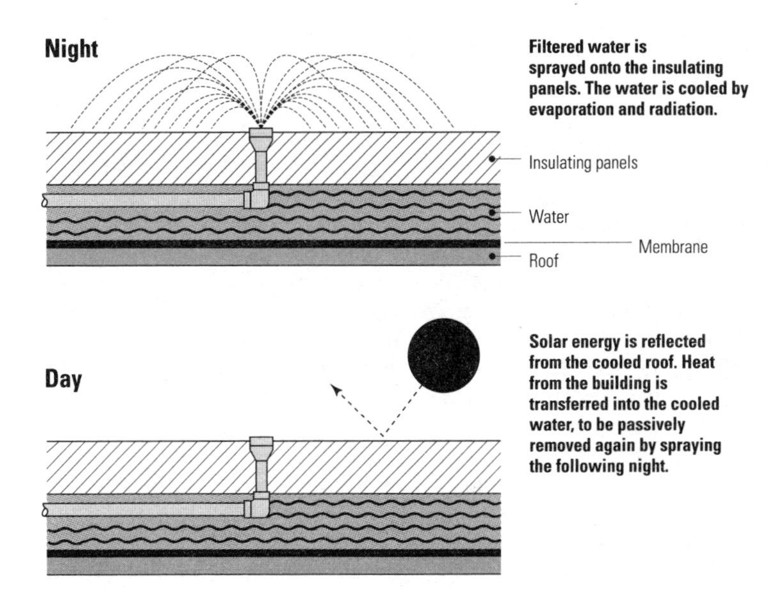

Figure 7 WhiteCap cooling can save as much as 90 per cent of cooling electricity. The device on the picture is installed in a house in Davis, California

Alternative Cooling

Three main alternative methods of cooling can do all the rest of the job anywhere in the world. Absorption cooling and desiccant dehumidification both handle humidity; they are driven not by a spinning shaft but by heat from fuel burning, an electric generator, an industrial process, or a solar collector. Evaporative cooling can deliver cool air, either moist or dry, into a space at quite low temperatures; well designed and driven by a small fan, it consumes modest amounts of water but very little energy.

Combinations of these three methods are especially effective. For example, a desiccant can first dry and warm the air, even in a humid climate. A direct evaporative chiller can then cool the air by evaporating water into it. In a heat exchanger, that cool, moist air can then produce cool, dry air. If it's dry enough, a little more water can be evaporated into it to make it even colder but not uncomfortably moist. Or in another combination, a gas-fired absorption chiller's waste heat can run a desiccant that makes it even more efficient and effective.

The first ACT^2 experiment retrofitted a 20,430-sq ft (1,900-sq m) section of the PG&E research offices in San Ramon, California, with efficient lights, some slightly improved office equipment and windows, and modestly improved insulation and draught proofing. This cut the cooling load in half. The entire cooling system was then replaced with an indirect evaporative chiller, backed up 5–10 per cent of the time by a very small, specially designed, very efficient refrigerative chiller. The design efficiency was probably a world record: only 0.14 kW/t, or 25 units of cooling delivered for each unit of electricity consumed. The 28-year-old Australian engineer who designed it believes he can do better next time. Monitoring data aren't yet in, but comfort has markedly improved. Changing from the original 2.0-kW/t rooftop units to 0.14 kW/t is a 93 per cent reduction in energy per unit of cooling. Since the amount of cooling required was also halved, the total reduction in cooling energy designed for the 'Sunset building' was about 97 per cent. As the older office equipment is gradually replaced by more efficient models, the one-half reduction in cooling loads will rise to about two-thirds, increasing the approximate cooling-energy saving from 97 per cent to 98 per cent – in a climate that can exceed 100F° (38C°).

SUPEREFFICIENT REFRIGERATIVE AIR-CONDITIONING

After the first four steps, conventional air-conditioning would no longer be necessary for comfort anywhere in the world. But if it were used, it could be made severalfold more efficient (see p 53) at reduced capital cost.

In a 1994 ACT^2 project, the California State Automobile Association built in Antioch perhaps the most efficient new office in California, saving three-quarters of total energy allowed by the nation's strictest energy standard. Yet comfort and amenity are exceptional, and it's the cheapest office CSAA has ever built. The air-conditioner was 40 per cent more efficient than a normal rooftop unit, and even better at part-load, but basically conventional. The cooling load was also cut about twofold through daylighting, superwindows and more efficient lighting and office equipment. The previous designers wouldn't accept the most efficient concept proposed (WhiteCap plus skylights); the design finally built was only the third-best idea. The best would probably have saved well over 90 per cent and cost even less. But the 72+ per cent achieved wasn't bad.

Controls and Storage

Whatever design is used, if it's not completely passive it will need controls. Better controls and software can typically save another 10–30 per cent of the remaining energy use. Control savings can even rise to about 50 per cent with careful training of the building operators on a computer simulator, analogous to those used to train airline pilots: big buildings are far too complex for operators to understand intuitively without such help.

Sometimes storing chilled water or ice can also save more energy – if by then there's anything left to store. It certainly saves electricity at peak periods, when most utilities charge more for it.

Multiplying the Savings

Successive energy savings don't add; they multiply. Each saving leaves less energy to be saved by later measures. But the savings do accumulate rapidly. Suppose, for example, that you save:

- 70 per cent of the amount of cooling required, by retrofitting better insulation, windows, lights and so forth (about two-thirds, which ranges from realistic to conservative);
- 20 per cent of the cooling need by expanding the conditions in which people feel comfortable (a reasonable and often low estimate);
- 80 per cent of the energy per ton of cooling by passive or alternative methods (remember the 93 per cent saving in the Sunset building);
- 50 per cent of the energy per tonne in the remaining refrigerative cooling (assuming any is still needed) – this term is normally more like 60–80 per cent;
- 20 per cent from better controls (normally the low end of the range) and nothing from storage.

Assuming an optimal constellation, the total saving could reach:

$$(1 - 0.7) \times (1 - 0.2) \times (1 - 0.8) \times (1 - 0.5) \times (1 - 0.2) = 0.0192$$

which means your cooling energy is now only 2 per cent of what you started with. That's how 'chains' of successive savings work: you don't have to save much at each step in order to get the total savings to multiply nicely, because there are so many steps.

Why doesn't everyone do it that way? Good question. We'll return to it in Chapter 4, p 146.

1.19 QUADRUPLING ENERGY PRODUCTIVITY IN FIVE SMALL STEPS

In the real world, achieving a factor of four in one big step can sometimes be too difficult. But there may still be scope for doing it by combining several smaller steps. This can be demonstrated by a simple example:

- Let us begin at the power-station end. A new generation of power stations, using the so-called combined-cycle gas turbines, can increase total efficiency to at least 50–55 per cent from the 34–40 per cent typical of classical steam power plants. (The latest combined-cycle gas stations are around 60 per cent efficient, with 65 per cent in view, but let's use the lower figure.) This increase means that 28 per cent less fuel must be burnt in the power station to produce 1 kWh.
- Adding combined heat and power and optimised gas boilers, a further 25 per cent can be gained on average for the typical mix of electricity and heat demand. Seventy-five per cent of the earlier demand then remains.
- Using fairly conservative measures for better insulation and more efficient appliances can provide a further 33 per cent of improvements, reducing the original demand from 100 per cent to 67 per cent. (To accommodate the conservative average citizen's habits, we are actively ignoring all that has been said before about Factor Four improvements both in insulation and in more efficient machines.)
- Furthermore, we assume that typical households can manage to reduce wastage of energy by a modest 7 per cent and accept another 3 per cent in reduced energy services through better controls that don't actually give up any desired amenities (*e.g.* less overheating, or turning off – manually or automatically – lights or fans or heating when leaving rooms for more than a few minutes). These minor improvements reap a further 10 per cent reduction.
- Finally, we assume that renewables can supply an additional 20 per cent to the energy mix, *e.g.* by passive solar heating, biomass and biogas use, reactivated small hydropower, wind-power and some photovoltaic energy. Together, these sources,

many cost-effective today, would then reduce the demand for traditional energy supplies by 20 per cent.

Taken together, these modest changes would reduce the need for coal, nuclear and large-scale hydropower by not the sum but the *product* of the parts:

0.70 x 0.75 x 0.67 x 0.90 x 0.80 = 0.25

thus needing only about one-quarter of what the demand is today. Assuming determination on the part of both public and private sectors, this goal could be reached in a matter of some 30 years under West European conditions and perhaps 5 or 10 years less in East European conditions. In developing countries, the calculation would be different. It would have to be adjusted for rapidly growing demand in energy services and for warmer climates, but also for a less efficient stock of existing power plants and a wider availability of renewables.

1.20 Profitable Energy and Waste Savings in a Louisiana Factory

Ken Nelson is an engineer who formerly directed energy conservation for Dow USA and has long been helping Dow Chemical Company's 2,400-worker Louisiana Division save energy and reduce waste. Dow is one of the world's largest and most sophisticated chemical companies – a leader in a cut-throat industry noted for penny-pinching. Dow's competitors would hardly say that Dow is stupid or lazy. Yet Dow has made the astounding discovery that there are $10,000 and $100,000 bills lying all over its factory floors – and that the more of them it picks up, the more it finds.

For each of the 12 years between 1981 and 1993, Ken Nelson organised a contest among the Louisiana Division's staff, never going higher than supervisor level, to elicit suggested projects that save energy or reduce waste, that can pay for themselves within 1 year, and (initially) that cost under $200,000. Submissions were peer-reviewed, and the most promising and profitable ones were implemented. After more than a thousand projects, shown by subsequent audit to have saved, on average, within 1 per cent of the predicted amount, some startling findings have emerged:

- In one of the 12 years, the average return on investment for the implemented projects slipped down to only double digits, at 97 per cent per year. But the other 11 years yielded average triple-digit returns, and for all 12 years, the confirmed return on the 575 projects that were subsequently audited averaged 204 per cent per year (*vs.* 202 per cent predicted), with total savings of $110 million per year (Nelson, 1993).
- In recent years, the energy savings being achieved became both larger and more profitable. Far from exhausting the cheapest opportunities, Nelson's contests were expanding the opportunities even faster through institutional learning and better technologies. It's as if each $100,000 bill picked up exposed a couple of new ones underneath.
- In the first year, 27 projects costing a total of $1.7 million had an average ROI of 173 per cent, and 'many people felt there couldn't be others with such high returns.' They were wrong. The next year, 32 projects costing a total of $2.2 million averaged a 340 per cent ROI. Learning quickly, Nelson changed the rules to eliminate the $200,000 limit – with such lucrative opportunities, why stick to the small ones? – and to include projects that would raise manufacturing output. In 1989, 64 projects costing $7.5 million saved the company $37 million in the first year and every year thereafter, for a 470 per cent ROI (the best so far). Even in the tenth year of the contest, 1992, nearly 700 projects later, the 109 winning projects averaged a 305 per cent ROI, and in 1993, 140 projects averaged a 298 per cent ROI.
- This manna from heaven was being picked up by ordinary workers, for no special reward except the recognition of their peers, and not because of the CEO's intervention, but because the CEO didn't know about it and therefore couldn't get in the way. Though meticulously measured and documented, Nelson's additions to Dow's bottom line didn't come from fancy management theories, quality circles, empowerment processes, committees or other managerial rituals; rather, they came from a practical shop-floor process that translated volunteer ingenuity into saved money. That is how markets really work, when they really work – and yet there are so few Ken Nelsons to make them work. How many market economists does it take to screw in a compact fluorescent light-bulb? None (goes the joke) – the free market will do it. But without a Ken Nelson and without the common sense and hard work of the workers he organised, the lamp might never get from the shelf into the socket.

It is not easy to estimate the total percentage of energy and waste savings from Ken Nelson's 12 years of devoted practice, or from similar efforts elsewhere that offer dozens of examples of quadrupling productivity over the years. You might expect such commercial success to be quickly emulated everywhere. Oddly, that does not seem to be happening. Even Dow's own Texas Division was resisting calls to imitate the adjacent Louisiana Division, because it has its own ways of doing things and, by some accounts, a classic 'not-invented-here' resistance to innovation. Indeed, after Ken Nelson retired in 1993, his organising committee was disbanded in a restructuring, tracking of further progress ceased and it therefore became impossible to evaluate how much progress, if any, continued without him. This illustrates the difference between demonstrated (let alone theoretical) potential and actual realization – an important distinction to which we shall return in Part II when considering market failures and practical ways to overcome them.

Chapter 2
Twenty Examples of Revolutionising Material Productivity

INTRODUCTION

The productivity of using material resources is not a familiar concept. We owe it mostly to Friedrich Schmidt-Bleek, the director of the Wuppertal Institute's Division for Material Flows and Eco-Restructuring. Schmidt-Bleek has developed the material inputs per service unit (MIPS) concept, a way to determine or estimate for any well-defined service the kilograms or tonnes of materials that must be moved about somewhere in the world. For a given service, material inputs might include the tailings from a copper mine in Chile, plus the water and other materials used and moved in the manufacturing process in Mexico, plus the packaging done in Chicago, plus some materials used and moved in the final sales process. Chapter 9 explains the MIPS concept in more detail.

Material productivity, then, refers to the reduction of MIPS. Longevity of products obviously helps to increase material productivity, where the services rendered are not significantly compromised as time passes. Think of old furniture which may actually gain in value over time. On the other hand, longevity stands in conflict with modernity, fashion and technical performance (including efficiency). Material productivity is a wider notion than durability; it refers to the product life cycle 'from cradle to grave' – or 'from cradle to cradle'.

Schmidt-Bleek (1994) feels that a factor of four MIPS reduction will not be enough. A factor of ten (see 'The Factor Ten Club', Chapter 9) is what he believes is necessary for the OECD (Organisation for Economic Co-operation and Development) countries. We hope our friend will excuse our pusillanimity when we nevertheless dare to include many examples showing a mere factor of four. Let's agree to call them a worthy beginning.

The Product Life Institute in Geneva, directed by Walter Stahel, has developed strategies for optimising resource efficiency. The context is the 'service economy' in which it's the end-use service that counts (Giarini and Stahel, 1993). The following elements may be of use for such a strategy:

- leasing instead of selling, wherein the manufacturer's interest lies in durability;
- extended product liability, which could induce manufacturers to guarantee low-pollution use and easy reuse or disposal;
- joint ownership or use (*e.g.* of cars or appliances), which would require fewer products for the same amount of services;
- remanufacturing – preserving the stable frame of a product after use and replacing only worn-out parts; and
- product design optimised for durability, remanufacturing and recycling.

Clearly, these elements imply a complex, multitask strategy. The material flows resulting from performing a task depend on how much of the task we do; how efficiently we use the materials required to do the task; how much ore has to be extracted and processed to obtain the materials needed; how far everything has to be shipped; and how many movements of material were involved in earlier years to create the infrastructure, the factories and the distribution networks. Material efficiency at every stage of the process determines, in a cumulative fashion, how much we can do with how little.

If everyone in our society is to have one widget, how many widgets do we need to make each year? Just enough to keep up with the number that break, wear out or are sent away, plus however many we need to keep up with net growth of the human population. The key variable is clearly how long the widgets last. To have something to drink out of, we need a lot fewer ceramic mugs than paper or Styrofoam cups, because the ceramic lasts almost forever (unless we drop it), whereas the 'consumer ephemerals' are used just once or twice and then thrown away before they fall apart. And if we make the ceramic mug unbreakable, it lasts long enough to hand on to our great-grandchildren. Once enough such unbreakable mugs are made for everybody to have one (or enough), very few need be made each year to keep everyone perpetually supplied.

2.1 DURABLE OFFICE FURNITURE

Durability is one of the most obvious strategies for reducing waste and increasing material productivity. Some parts will wear out faster than others or lose their aesthetic appeal for reasons of wear, fashion or a change in corporate identity. If these parts are made interchangeable, much can be gained in increasing durability.

Let us look at an office chair, an example studied by Walter Stahel (to whom we owe much of the substance of this section). One of the pioneers of durability thinking, Stahel suggests that the most promising technical approach to combat early disposal is to separate 'structural elements' from 'visible elements'. Combined with the marketing approach of an availability of upgrades for the visible elements and a complete product take-back by the manufacturer, the 'eternal' office chair may not be far in the future: indeed, the second-largest manufacturer of office furniture in America, Herman Miller, has opened a plant specifically designed to remanufacture indefinitely every kind of furniture it has ever made, and this 'Phoenix' programme is proving highly successful.

The visible and exchangeable elements can be 'dematerialised' (produced with minimal MIPS) and designed for easy remanufacturing or recycling. Visible and structural elements are clipped together and easily declipped again when the time comes for exchanging the worn-out elements. The structural elements of an office chair include its 'foot', its 'leg' and the mechanics of the seat. These structural basics can be optimised for top ergonomic quality, comfort, robustness, durability and easy repair. And yet they leave plenty of room to apply fashion creativity to the cushion. If owners change or the company chooses a new design, wants a new corporate identity or just wants to give personnel the feeling of using brand-new office furniture, exchanging the cushion and the cloth is a minor matter both financially and ecologically.

Many famous furniture products have been built to separate structural from visible elements. Furniture museums are proud to exhibit chairs by Le Corbusier or Eames that were designed in this way, although they were not mass-produced. The principle entered the mass-manufacture markets in Germany when the government legislated obligatory take-back of durable goods. Furniture manufacturers such as Sedus, Wilkhahn and Grammer have started to apply the above principles to their new collections.

Data are not readily available for a clear-cut proof of attaining a factor of four in MIPS reduction by this method. However, a superficial estimate suggests that factors between 5 and 20 can be assumed, depending, of course, on the materials used and on the basis for comparison.

Long-lived office furniture could theoretically become a nightmare for the manufacturing industry. Market saturation could soon be reached. Exchanging cushions and upholstering chairs would become a gainful activity for local workshops rather than for the manufacturer. The most promising strategy for manufacturers could be leasing instead of selling the furniture, which could put a premium on their maximising durability. Such an inconspicuous change could have a fundamental effect on the structure of the whole industrial economy – it might be the starting signal towards a utilisation-focused service economy.

2.2 Low-MIPS Car/Hypercar

Driving an automobile means more than moving one's body about. You also move a tonne or more of car about. Moreover, according to the MIPS philosophy, a lot of material weight is moved about just to manufacture the car in the first place. Friedrich Schmidt-Bleek's group estimates that more than 1,520 tonnes are moved in the sequential processes of metal mining, refining, shipping, plastics and glass manufacturing, and assembling the car. The catalytic converter alone, weighing a mere 9 kg but containing a few grams of platinum, induces material flows of more than 2 tonnes because so much rock must be moved and ore processed to produce each speck of platinum. But by redesigning an auto for low material intensity, higher service life and lower fuel use, a factor of four can be won in MIPS reduction.

Hypercars

A far more radical strategy is the total redesign of the car along the lines of the 'hypercar' design philosophy (see p 4).

In the US, the automobile industry and associated sectors represent directly and indirectly about one-tenth of total employment and consumer spending, and one-seventh of GNP (Gross National Product). They use approximately 70 per cent of the lead, about 60 per cent of the rubber, carpeting, and malleable iron, 40 per cent of the machine tools and platinum, 34 per cent of the total iron, 20 per cent or so of the aluminium, zinc, glass and semiconductors, 14 per cent of the steel and 10 per cent of the copper. Automotive use of materials has changed only slowly in recent decades: during model years 1984–94, for example, the average US production car got 1 per cent heavier and shifted its mass composition only three percentage points, chiefly from steel to non-ferrous metals and polymers. But with the advent of ultralight-hybrid hypercars, most of the auto industry's immense material flows would change rapidly and profoundly.

Hypercars will soon come to weigh about threefold less than today's steel production cars, and will shift essentially all structural mass from steel and other metals to polymer composites. According to a detailed proprietary study recently completed by Rocky Mountain Institute (Lovins *et al.*, 1996), even a very early, illustrative and unoptimised four- to five-passenger hypercar design using a water-cooled 20-kW (15-hp) external-combustion engine, a 50-kg nickel-metal-hydride buffer battery, glass glazings, refrigerative air-conditioning and other off-the-shelf technologies could readily weigh two-thirds less than an average 1994 US production car: conservatively, 521 vs 1,439 kg kerb mass. RMI's mass budget with 110 line items, based on benchmarking against existing products and prototypes, suggests that such a hypercar, compared with the average 1994 US production car, could contain approximately:

- ❑ twice the mass of composites and other polymers,
- ❑ one-eighth more copper,
- ❑ 92 per cent less iron and steel,
- ❑ one-third less aluminium,
- ❑ three-fifths less rubber, and
- ❑ four-fifths less platinum and nonfuel fluids.

Such an early design, however, is close to a maximum-metals case. Alternative electric buffer storage and powerplants, both expected to become widely useable in the late 1990s, would eliminate about three-fifths of the remaining metals, including all the iron, nickel and metal-hydride alloy, perhaps half the aluminium and much of the steel. (Iron and steel use could then drop by 96 per cent or more, not 92 per cent, from today's production cars.) These and other refinements could also reduce the kerb mass towards closer to 400 kg. Copper use would moderate and could eventually become about the same as in today's cars. Platinum use could return to about today's levels only if the onboard powerplant used certain kinds of fuel cells. Some specialised metals such as magnesium and titanium might gain small but useful niche markets, but on the whole, the design would massively displace metals with sophisticated polymers.

Would the increase in advanced composites – plastic resins reinforced by such ultrastrong materials as carbon fibre – mean a great expansion in the plastics industry? Not at all. Cars today use 7 per cent of OECD, 5 per cent of world and 3 per cent of US polymer production. Moreover, polymers and composites already constitute 8 per cent (US) or 9 per cent (world average) of the average car's weight and perhaps 20–30+ per cent of its materials volume. However, that 8 per cent of an automotive mass as heavy as today's averages 111 kg of plastics and composites. That already exceeds the likely total mass, around 100 kg, of a hypercar's structure, body and closures (doors, boot, and bonnet), excluding all attached components such as interiors, trim and driveline. It's also more than half of the approximately 227-kg total mass of polymers and composites used throughout an early hypercar. For this reason, converting the entire US auto industry to hypercars would increase the total US use of polymers by only 3 per cent, less than 1 year's normal growth. However, it would mean an order of magnitude expansion in the advanced composites industry (whose annual turnover in 1995 was about $10 thousand million) and a several-hundredfold expansion in the currently small production of carbon fibre.

About two-thirds to three-quarters of the mass of the hypercar will be components added to the composite body-in-white (*vs* approximately 81 per cent in today's threefold-heavier cars). Most hypercar components will be similar to today's, but much smaller and lighter. However, many components will disappear altogether with the elimination of such commonplace elements as power steering, power brakes, axles, transmission, clutch, differentials, alternator, starter motor and so forth. The powerplant could initially be an internal-combustion engine approximately 10–25-fold smaller than today's, but would probably soon change to other kinds, ranging from modestly different (Stirling or gas-turbine) to profoundly different with no moving parts (fuel cell or thermophotovoltaic). The electrical buffer storage device will initially be a nontoxic, recyclable nickel-metal-hydride battery about three times heavier than the ordinary approximately 14-kg lead-acid starting battery in today's cars, but will then probably soon change to a carbon-fibre superflywheel (a precursor of which, an offshoot of Britain's centrifuges for enriching uranium, entered the market in late 1995) or an ultracapacitor weighing about 10–20 kg, or perhaps even a thin-film lithium battery weighing as little as 5 kg. In any event, it will contain no lead and very little metal of any kind.

Hypercars would need an order of magnitude less consumable fluids than today's cars (the main remaining ones would be fuel, reduced by about tenfold, and windscreen-washing fluid). Mature hypercars would eliminate 6–8 of 14 kinds of fluids needed by today's cars and often partly or wholly lost to the environment. In particular, motor oil, with its benzene, heavy-metal and other troublesome contaminants, would be greatly reduced or eliminated – a major benefit, since an average US car uses 22 litres per year. Also greatly reduced or eliminated would be fuel and oil additives (for engine cleaning, life extension or cold-weather starting), distilled battery water, antifreeze (plus radiator water and anticorrosion additives or flushing agents on occasion), brake fluid, power steering fluid, automatic transmission fluid, grease, various liquid and solid lubricants, and air-conditioner refrigerant.

Some 12–13 of the 21 main categories of regularly replaced mechanical components would also disappear or would last as long as the car, while the rest would greatly decrease in size and in frequency of replacement. The material flows that would be reduced or eliminated, depending on details of the powerplant used, would include drivebelts (for radiator fan and water pump, alternator, air-conditioner compressor, power-steering pump, exhaust-gas-recirculation air pump, etc.); hoses (for air, coolant, fuel, oil, refrigerant and vacuum); starter batteries; clutch components; timing belts; light-bulbs (scores per car); brake shoes and pads; air filters; oil filters; and spark plugs. Furthermore, the flow of spare parts generally would be substantially reduced, as would the frequency, extent, and pollution of body repairs.

As for the cars themselves, each year in North America, more than 10 million cars, or 94 per cent of all cars disposed of, are dismantled, then three-quarters recycled (contributing 37 per cent of US steel scrap), and one-quarter landfilled as hetereogeneous and somewhat toxic 'shredder fluff' (which typically includes 42 per cent fibre and 19 per cent polymer). The recycled metals are equivalent in tonnage to about all of the steel and one-third of the nonferrous metals used each year to make new cars (although the recycled steel does not go back into new cars; in practice, it's diluted with other scrap less contaminated with copper, then reused mainly in structural applications). Automotive plastics are not yet generally recyclable, although German- and Swedish-led innovations that reduce the number and improve the labelling of polymers may change this situation in North America, as they have in Europe. However, hypercars would eliminate the steel body and most other metallic parts whose scrap value now finances recycling.

Except for the dismantlement stage, and that only with retraining of employees, hypercars cannot be processed by the current car-recycling infrastructure. This is not an immediate concern to car recyclers, since they should keep busy for at least another couple of decades disposing of steel cars. If, however, shredder fluff were declared hazardous waste, its high disposal costs would exceed the value of recovered metals and thus immediately destroy the car-recycling industry.

Hypercars, on the other hand, do offer novel kinds of attractive recycling options, in successive stages:

- life extension and even 'reincarnation' over decades for different markets, perhaps even in different societies, taking advantage of the cars' software, changeable colour-skin, and other options for upgrades and 'personality changes';
- extensive reuse and remanufacturing;
- primary recycling that recovers the valuable composite fibres (currently through solvolysis, notably methanolysis – decomposing the resin by exposure to pressurised and heated methanol so that the valuable fibres can be recovered for reuse – with several other options emerging); and
- failing all else, secondary recycling by chopping or grinding into valuable filler material; or
- tertiary recycling by pyrolysis to recover energy content and molecular building blocks.

The highest-value processes appear robustly cost-effective and are now commercial, although they can be considerably refined and need long-fibre recovery techniques for this new application. Tertiary and probably secondary recycling, which destroy the valuable long fibres, should seldom if ever be needed.

Moreover, even if every car were a hypercar and lasted no longer than steel cars, and the entire composite and polymer mass of every hypercar were landfilled rather than using available recycling options, the resulting mass of thrown-away polymers and composites would still be less than the 331 kg of shredder fluff being landfilled in North America today (or about three times that fluff's polymer mass) – but unlike fluff, would be essentially non-toxic. For more information please refer to Lovins *et al.*, 1996.

2.3 ELECTRONIC BOOKS AND CATALOGUES

Physicians like the *Merck Manual.* It's the world's leading medical reference book, some 3,000 pages thick. The trouble is that you can't routinely carry so much paper when you're visiting patients at their homes, especially if they live on the fourth floor of an old-fashioned townhouse without a lift.

Now, in the electronic age, there is an alternative: the *Merck Manual* plus the *Physician's Desk Reference*, all on a palm-size CD-ROM. With a suitable CD-ROM–adapted laptop computer, doctors can now consult both reference books while sitting at their patients' bedsides. What a relief for them, and what an improvement in material efficiency!

Similarly, *The John Haldeman Diaries: Inside the Nixon White House* (the 700-page book) plus 2,000 more diary pages along with 700 photos and 45 minutes of movie footage shot by the late chief of staff are offered on the market on one CD-ROM for $70. This price is about as much as the material, printing and shipping costs of the same stuff in the form of paper and film, weighing more than 5 kg.

Electronic publishing is advantageous for more than books. Newspapers can come via e-mail and be read on-screen at home or on the ride to work (unless you have to steer your car). Articles and news worth saving can be marked for printing and/or electronic filing. Not only does electronic filing save material resources, it's also much more convenient and reliable for retrieving than today's clipping and paper filing (see Plate 7).

REPLACING PAPER CATALOGUES WITH BITS AND BYTES

A typical architectural or engineering office contains whole rooms lined with vast shelves full of heavy, awkward parts catalogues. Every manufacturer must bear the cost of preparing, printing and shipping these monsters, typically once a year, thereby consuming untold forests and oil resources. And the design professionals in turn must spend much of their time leafing through thousands of pages for just the part they want – then laboriously copy and redraw it for entry into the electronic plans for their projects. A *Wall Street Journal* story reports that 'Many designers spend 16 to

20 hours a month copying [and scanning] the parts drawings in these books, so that they can be edited and resized on their [personal computers].'

The oddest feature of this antiquated system is that some 2.5 million of these same customers *already* use a single software program, called AutoCad, to design everything from machines to buildings; and over 80 per cent of these designers must specify parts from manufacturers, many of whom use that same software to design the parts in the first place. An obvious opportunity to save resources and time.

By customer demand, AutoDesk (San Rafael, California), the fifth-largest vendor of personal-computer software and the provider of AutoCad, has begun to publish those huge manufacturers' catalogues in digital form, so that the drawings and specifications for the parts can be electronically transferred directly into design drawings. For example, a $99 CD-ROM disk called PartSpec contains all the details of more than 200,000 parts from 16 leading manufacturers. Another, a $199 CD-ROM disk called MaterialSpec, includes more than 25,000 materials from 300 manufacturers.

Not all manufacturers welcome this system, because it helps customers directly compare competing products more easily than with paper catalogues. But AutoDesk used the clever stratagem of persuading one leading vendor from each parts category to post its materials on the first edition of its CD-ROMs. Customers, they reason, will so welcome this convenience that other vendors will have to follow suit – or risk making their products far less accessible to the designers whose specifications control their sales. 'Over time,' reports the *Journal*, 'AutoDesk expects to publish drawings from virtually all comers.'

The concept could easily be extended from, say, the gears, motors, pulleys, fans and other parts used by mechanical engineers to the wallpaper and furniture used by interior designers, or the windows and doors used by builders, or whatever your imagination suggests. Presumably it's then but a small step to electronic ordering, akin to existing online mail-order catalogues for household products.

'This has the potential,' the *Journal* quotes one industry expert as saying, 'to move many professionals away from paper catalogues.' Thus arises yet another reason for saving resources (in this case trees, papermaking energy and water, and transportation energy): the electronic product will be enormously faster and easier to use, being electronically searchable in an instant and then electronically pasteable into design drawings. This saved time, not to mention higher-quality, more error-free drawings, will yield benefits that go far beyond saving material resources. It may well turn out that the ease of comparing products will further improve competitiveness – partly by helping users see at a glance which products have been designed with the most elegant frugality (see Clark, 1995).

2.4 STEEL VERSUS CONCRETE

Buildings, bridges, pylons for high-voltage mains and many other structures can be made with steel or with other materials, notably wood or concrete. Although wood has been underestimated in its heavy-duty capacities (see p 108), it seems more promising for the purpose of a material efficiency assessment to compare steel with concrete. The 'steel group' at the Wuppertal Institute, chaired by Christa Liedtke, has made such comparisons using the MIPS concept. Gradually the group became ever more convinced that steel deserves a renaissance.

For a simple first-glance assessment, steel and concrete pylons for carrying 110-kV electric mains were selected and compared. Their service function can be described as 'carrying mains of 110 kV for a fixed time span of *e.g.* 40 years'. Given such a well-defined service function, the material turnover or input per service unit has been estimated (Liedtke *et al.*, 1993).

Two main differences make steel a more ecological choice under the MIPS concept:

- Concrete pylons require in their initial construction three times as much material as steel pylons. For a typical pylon in use in central Europe, the absolute figures are 90 tonnes of materials turnover for a concrete pylon (weighing itself some 45 tonnes). For a steel pylon performing the same function, the figures are 36 tonnes and 6 tonnes respectively.
- The service life of steel pylons can be twice or more as long as that of concrete pylons; regular maintenance is required, however, every 10 to 20 years, depending on climatic conditions.

Moreover, steel pylons can be constructed from scrap iron and steel, improving the balance even further in favour of steel. Another factor of 2.5 can be assumed to be gained.

All in all, a sixfold material efficiency can be stated for a transition from concrete to steel pylons, as is symbolised in Figure 8.

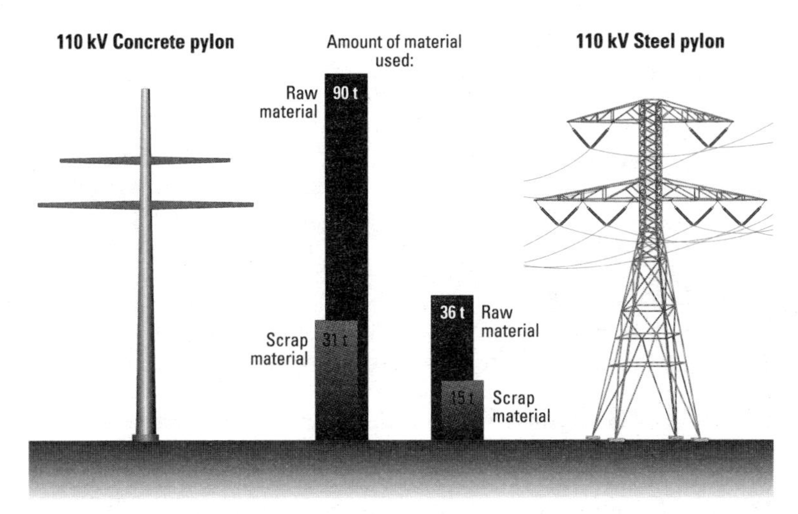

Figure 8 A MIPS comparison between a concrete pylon and a steel pylon of the same functional quality.

Traditionally, only steel was used for mains, pylons, bridges and so forth. But after World War II concrete came in forcefully in virtually all OECD countries, even those with a strong steel industry. A historical-economic assessment by Liedtke *et al.* seems to show that the reasons for replacing steel by concrete had little to do with costs, but rather with fashion or engineering 'schools'. Concrete was seen as more modern and more 'elegant'. With an avalanche of repair needs now appearing, however, concrete may be falling in favour relative to steel. The MIPS/Factor Four philosophy could easily accelerate this process.

The situation could further improve for steel if the efficiencies of manufacturing it could be optimised. Steel is still chiefly produced in the basic oxygen furnace. This method wastes energy, water and materials. Another method, electric smelting, has been introduced but is still far from replacing the old methods. Electric smelting involves much less mass of materials than the oxygenation method. Per tonne of steel, Liedtke *et al.* found that electric steel uses one-tenth of the fuel, one-eighth of the water, one-fifth of the air and less than one-fortieth of other materials compared with traditional basic-oxygen steel. On the other hand, some 30 per cent more electricity is needed for electric steel. If electricity is counted according to the ecological rucksack (cf p 242) incurred by the way it is typically generated in Germany, the balance is no longer so favourable. Still, a factor of four in improving material efficiency can be defended for the transition from oxygen steel to electric steel.

The commercial interest in long-lived structures, and therefore in steel, could be spurred if pylons, bridges and other such structures were leased instead of being sold. With leasing, the construction firm would have a vital commercial interest in durability and low maintenance costs. The time may actually be ideal for introducing the leasing concept for bridges. Many municipalities in Germany and elsewhere are highly indebted and would have to borrow money to build new bridges. Offering leasing contracts costing no more than capital costs plus maintenance costs for concrete bridges, which can be onerous, should be attractive – unless, of course, carbon fibre technologies, ever cheaper as new industries like hypercars grow (p 71), become even more attractive by virtue of their great resistance to corrosion and fatigue and their potential for snowballing weight savings.

2.5 SUBSURFACE DRIP IRRIGATION

Sundance Farms in Arizona's Casa Grande Valley is a model of efficiency in irrigated agriculture. On 2,000 acres (about 830 hectares) of cotton, wheat, barley, milo, maize, seedless watermelons, cantaloupes and sweet corn, Howard Wuertz is demonstrating that agricultural resource efficiency is a package deal.

For some resources and human activities, typical current efficiencies make Factor Four or greater increases in efficiency hard to achieve. In irrigated agriculture, while some farms are terribly inefficient in their water use, many larger commercial farms achieve water-use efficiencies in the 40–60 per cent range. That is, of all the water applied to a field, 40–60 per cent is taken up by the crops to satisfy their evapotranspiration requirements. The rest is lost to surface runoff, deep percolation or sprinkler wind spray. Improving water-use efficiency to 100 per cent, so that every drop of water applied to the field ended up evaporating from the crop itself, would increase resource efficiency by a factor of 'only' 1.7 to 2.5.

When Howard Wuertz began changing his fields from furrow and flood irrigation to subsurface drip irrigation in 1980, he increased field water-use efficiencies from roughly 60 per cent to over 95 per cent, a factor of 1.6 improvement. The drip lines, buried 8–10 inches (20–25 cm) deep, emit small amounts of water right in the plant root zone. The soil surface usually stays dry, reducing surface evaporation, and the root zone is never saturated, reducing runoff and deep percolation. The few per cent of water lost is mostly accounted for by the occasional backflushing of the drip lines.

The water savings were important in arid Arizona, but perhaps more important were other benefits. First, Wuertz found he could reduce tillage operations, replacing ploughing, floating, land planing and listing (all separate steps to prepare the seed-bed for planting and efficient surface irrigation) with simple shallow surface tillage. Studies by the University of Arizona on his farm showed that he had reduced tillage energy use by 50 per cent. Simplified tillage also allowed quicker postharvest turnaround of fields, permitting two crops to be harvested in some years. Next, because the drip lines cut water losses, less of the applied herbicides and fertilisers left the fields. Herbicide applications were reduced by 50 per cent and nitrogen fertiliser use by 25–50 per cent. Also, less water had to be pumped from deep well turbines, thereby reducing pumping energy use by 50 per cent.

Finally, crop yields increased by 15–50 per cent. A variety of factors probably contributed: greater uniformity of water application, greater effectiveness of systemic insecticides now delivered through the drip lines directly to the plant roots, and better management of yield-reducing salts that often accumulate in surface-irrigated fields. Higher yields with less water meant a reduction in water use by a factor of 1.8 to 2.4 – in a hot and unforgiving desert where rising water costs had already wrung out the most obvious savings.

Sundance Farms is no hippie organic vegetable stand. It is a serious commercial production operation. The drip lines, made to last and buried below the depth disturbed by any agricultural equipment, were dear to install, but the cumulative reductions in inputs and increases in productivity made the investment very cost-effective. In coming years, following the lead of progressive farmers like Howard Wuertz, savvy mainstream agribusinesses will increasingly seek the multiple benefits of improved agricultural technologies and management practices.

And to those looking for Factor Four or more improvement, we say either change to less water-intensive crops (cotton was hardly designed to grow in deserts) or use the crops themselves more efficiently.

Who, you might wonder, invented the concept of dispensing precious water to desert crops a drop at a time, directly to the roots, and only as fast as the plant required? Why, the Anasazi – the long-gone 'ancient ones' of the American Southwest. They would bury an unglazed earthenware pot up to its neck in the ground, fill it with water, put a lid on it, and plant maize and beans right round it. Fed by the slow seepage of water through the clay pot, the plants would grow up round it, their roots growing over and into the moist clay, and their leaves shading the lid from the sun. Every few days or a week, another pot full of water would be added. Today we have high-tech polymer emitters and tubes instead of clay pots, and computer controls instead of children to keep up the water supply, but the modern drip system's mode of operation is remarkably similar.

2.6 WATER EFFICIENCY IN MANUFACTURING

PAPER AND BOARD

In 1900, paper manufacturers in Europe would typically use a tonne of water per kilogramme of paper produced. By 1990, the relation improved more than 15-fold downwards to 64 kg of water used for 1 kg of paper and board. Of this, 34 kg went into pulp production, and 30 kg into manufacturing paper and board from pulp (Liedtke, 1993).

Further improvements down to some 20 or 30 kg of water were made more recently in Germany, chiefly as a result of rising wastewater charges. And some manufacturers went to the extreme. A North German paper manufacturer has managed to avoid wastewater altogether for the production of paper for packaging purposes. All water from the manufacturing process is collected and filtered for reuse and is only supplemented by tiny amounts of fresh water to provide the water molecules needed for the mechano-chemical stability of paper and to make up for evaporation. This plant has ended up using no more than 1.5 kg of fresh water per kilogram of packaging paper. Figure 9 shows the splendid progress in reducing water use.

Total recycling of water is achieved by consecutive processes of sedimentation, flotation and filtration of the particles contained in the water used both in the pulp-production and the paper-manufacturing cycles.

In the 'factor language' of this book, the achievement by this manufacturer is roughly a factor of 20 against the current European average and, for packaging paper, a further factor of 8 against what the same factory had achieved before its latest efficiency innova-

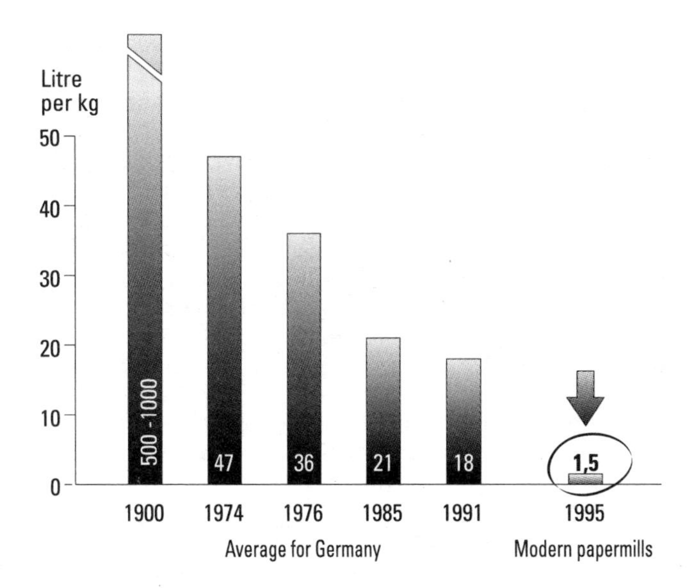

Figure 9 The triumphal progress of reducing water use in paper manufacturing. The graph shows the 40-fold decrease (in the case of paper for packaging purposes even 600-fold) of the average freshwater consumption of paper and board manufacturing from 1900 to 1995.

tion. The present state, however, is arguably the end of the road, because some water must go into the chemistry of the paper, and the recovery of water from vapour would incur excessive costs.

Razor Blades, Pens, and Microchips*

Gillette Company, one of the world's leading manufacturers of shaving equipment, used 96 per cent less water to make a razor blade in its South Boston Manufacturing Centre in 1993 than in 1972. Gillette's Santa Monica Manufacturing Centre also used 90 per cent less water to make a Papermate pen in 1993 than in 1974.

Were these huge water savings achieved through quantified goals set by top managers? No. The managers simply said they wanted resource efficiency to be a priority. This encouraged

* We thank Wendy Pratt of California Futures, Sacramento CA, for communicating the data used for this subchapter. The data originate from Gillette Company USA and Mitsubishi Semiconductors America.

employees to look continually for ways to improve production processes. Teams of employees at every Gillette site strove to elicit and implement ideas that work together to yield results no single individual could have achieved.

At one plant, an employee suggested recycling process water, but that would have required building a new cooling system. So another employee suggested using the swimming pool as the 'cooling system'. These ideas were then shared with Gillette's European sites, whose employees added still more refinements. Eventually, the Boston plant came to get all the heat it needed from the heat it had formerly been venting.

Coupled with these resource savings were process improvements that reduced pollution far beyond the expectations of the Environmental Protection Agency's 33/50 programme. For example, actual releases of 1,1,1-trichloroethane, trichloroethylene and methyl ethyl ketone (three common but disagreeable solvents) were reduced by 98 per cent from their 1991 levels as washing of blades and other parts was switched over to water-based systems. By 1993, Gillette had reduced its Toxic Release Inventory wastes by 97 per cent from their 1987 level.

Such dramatic improvements are available even in the most sophisticated high-tech manufacturing, where one might suppose opportunities for substituting materials are most limited. For example, starting in 1991, Mitsubishi Semiconductors America started reducing water use and waste production in its North Carolina integrated-circuit factory. While production increased 30 per cent, water use fell by 70 per cent, and industrial wastewater sludge output by 75 per cent (from 1992 levels). Treatment chemicals were also saved, water quality improved, and productivity rose. The $1.2-million cost of the two main improvements – deionization and reverse osmosis – paid back in 2 years; indeed, the first phase of water savings cost $40,000 and saved $240,000 in the first year (a 9-week payback).

Moreover, hazardous waste production has fallen by 75 per cent overall since 1991, including complete elimination of CFCs and a 96 per cent drop in plating sludge wastes (by changing system chemistry with no change in production). An electrowinning process, for example, reduced the lead concentration in plating sludge from 1,100 to 30 parts per million, a 97 per cent reduction, further reducing chemical reactant use, further reducing sludge generation.

2.7 Residential Water Efficiency*

Residents of a typically water-wasteful household in the US consume approximately 80 US gallons (303 litres) of water per person per day (gpcd) for indoor use alone. Meeting the US National Plumbing Standards that went into effect in 1992 could reduce that figure by 35 per cent to approximately 50 gpcd simply by replacing old plumbing fixtures with the worst permitted under the new law – 1.5-gallon-per-flush toilets and 2.5-gallon-per-minute showerheads and taps. Using instead a toilet that meets the standards in most parts of Australia (a dual-flush 1.5-US gallon-per-flush toilet), more efficient dish- and clothes-washing machines, and greywater for toilet flushing could reduce that 50 gpcd by another 50 per cent, or by a total of 69 per cent, and provide over 30 gallons per day of greywater for outdoor needs.

The technologies required look similar or identical, work at least as well and needn't even cost more. The Australian dual-flush toilet represents an 80 per cent reduction over the once standard 5-gallon-per-flush toilet, yet works better and costs about the same (sometimes less). A Swedish 0.79-gallon-per-flush toilet used by many of Rocky Mountain Institute's more than 40,000 visitors since 1983 has proven highly reliable; such designs provide better transport than the old 5-gallon-per-flush models because their 84 per cent lower volume of water, instead of swirling round aimlessly, is concentrated into a strong, precisely shaped pulse. Similarly, the top-loading, horizontal-axis washing machine developed by Staber Industries, a small firm in Groveport, Ohio, uses a special geometry to swirl the water severalfold more powerfully through the clothes, getting them cleaner with only about one-third as much water (and water-heating energy) and one-quarter as much soap as usual.

Still larger savings are available from specially designed equipment. Around 1980, for example, an American inventor marketed a powerful Min-Use shower whose frugal 0.5-gallon-per-minute water spray was vigorously propelled by a blast of low-pressure warm air – a derivative of the Buckminster Fuller concept of mist showers for submarines. Using four to five times less water than a fairly efficient shower head, or 8–15 times less than a pre–1992 model, the Min-Use took only 1–2 per cent as much power to run the blower as was saved by heating less water: 0.43 kW, compared with 20–75 kW. The extra cost of the equipment was largely offset

* We owe the draft for this subchapter to Scott Chaplin, RMI.

by much lower installation cost, since a very small, flexible pipe – not a big, rigid one requiring a plumber to install – could carry all the hot water. In fact, the pipe was so small that it shouldn't be insulated: any but very thick insulation would waste energy by increasing the pipe's total surface area more than it reduced heat loss from copper to air.

Efficient use of water can often be combined with alternative water supplies. For example, using rainwater for all fixtures except taps would reduce public supply withdrawals by 90 per cent in many areas, save soap (because rainwater is so 'soft' that little is needed for cleaning), and also provide substantial amounts of greywater for outdoor use.

While use of greywater and roof-captured rainwater may sound extreme to many, these sources are now being used in modern systems round the world. In many parts of Hawaii, the Caribbean, Australia and even some parts of Texas, people rely on rainwater for most, and often for all, of their water needs. In Germany, rainwater collection systems are promoted as a method of reducing storm water runoff problems. In Tokyo, many new office buildings and the Tokyo Dome baseball stadium have rainwater collection systems to provide water for toilet flushing and cooling towers. Greywater systems have been gaining popularity in California since laws were passed allowing greywater use for subsurface irrigation or toilet flushing.

Outdoor water use can also be reduced significantly. In many parts of the US, outdoor water use is the same as or greater than indoor water use during dry months. The range of savings available is almost unlimited, depending on the user's landscaping needs. Efficient irrigation systems, limited turf areas and plants that consume less water can easily reduce outdoor water needs by 50 per cent or more in many areas. Use of greywater and rainwater can reduce the need to use municipal water. All these techniques can often bring savings in fuel, fertiliser, herbicide and labour. In addition, well-designed water-efficient landscapes can be aesthetically pleasing, provide natural cooling and fire protection, and attract a wide diversity of birds and other wildlife.

The Casa Del Agua pilot project in Tucson, Arizona, was an ordinary-looking, fully equipped, very pleasant home that used 67 per cent less municipally supplied water than average homes in the area. In this project, a combination of rainwater and greywater systems, water-efficient landscaping and water-efficient fixtures brought total per-capita use of municipal water down to approximately 50 gallons per day. A similar pilot project in Phoenix called the Desert House is now underway and is expected to achieve at least a 50 per cent reduction in water use compared with similar homes in that area. Both projects were developed by the Office of Arid Lands Studies at the University of Arizona College of Architecture (Martin M. Karpisack *et al.*, 1990).

Water efficiency may be improved not only in single buildings but also in whole communities. When Goleta, California, faced high costs for new water supplies during a drought, the water authority helped its 74,000 people install over 17,000 water-frugal toilets (14,700 of them aided by rebates), gave away about 35,000 high-performance showerheads, identified more efficient irrigation methods for hundreds of households and aligned water tariffs with marginal costs so that people would understand what their next unit of water usage really cost the community. From May 1989 to April 1990, per capita residential water use fell by over 50 per cent compared with the previous 5-year average. Total usage fell by over 30 per cent, from 135 to 90 gpcd – twice the 15 per cent target. The average single-family, multifamily and motel savings were respectively 50, 40, and 40–50 per cent. Later savings raised the total savings from over 30 per cent to 40 per cent. The whole programme, costing $1.5 million, reduced sewage flow from 6.7 to 3.9 million gallons per day by June 1990, indefinitely deferring a multi-million-dollar expansion of the previously overloaded treatment plant that had been thought necessary to comply with EPA standards (RMI, 1994).

Using recycling techniques originally developed for spacecraft, some American experimenters have provided ample water supplies to their families using no external water supplies, but literally reusing every drop. After all, that's what the earth does every day; it just takes some special equipment to do it on a smaller scale. On this principle, Ted Bakewell III of Ladue, Missouri, developed mobile homes that require no hook-ups to anything: they recycle all their water and wastes, capture all their energy with solar cells and a small wind rotor, provide their own wireless telecommunications and can be delivered by air, water (they're amphibious) or airdrop to wherever people want to live. No infrastructure: when you want to go live somewhere else, you just take it away again, leaving no scars or pipes behind.

Of course, to avoid producing any trash as well, you would need to be as careful in your purchasing, recycling and composting habits as the California engineer who hadn't carried out the rubbish for years – because he didn't need to: his net production of solid waste was less than 40 g a month!

Incidentally, some people wonder what the ultimate practical limit to recycling residential wastes is. In 1987, Barry Commoner of the Centre for the Biology of Natural Systems (CBNS) at Queens College, City University of New York, Flushing, New York, conducted a pilot project for an intensive residential recycling system in order to determine what percentage of residential rubbish could actually be recovered in marketable forms. Rubbish collected from volunteer residents was separated into compostable materials, recyclable materials and nonrecyclable materials. The compostables and the recyclables were sent to nearby processing centres. Only 2.4 per cent of the sorted materials set aside for recovery was rejected. Of the original residential rubbish collected, 84.4 per cent was recovered by composting or recycling.

2.8 COTTON PRODUCTION WITH LESS WATER

We all need clothes. Textiles are among the most important products of any civilisation. Thirty-seven million tonnes of textiles were produced worldwide in 1990. In the same year in industrialised countries roughly 20 kg of textiles were sold per person. In the developing countries the figure is only about 1 kg per person.

Textile manufacturing is the cause of many well-known environmental problems ranging from water pollution from the dyes to pesticides used to protect wool and unpleasant chemicals involved in the production of artificial fibres.

Less well known are environmental problems linked to the material flows that result from the manufacture of textiles. One such material flow sticks out as particularly nasty – the water consumption of cotton plantations, and related soil erosion. For every kilogram of cotton fibre produced, typically 5 tonnes of water are flowing. The worldwide production of 18 million tonnes of cotton annually thus implies the flow of some 100 thousand million tonnes of water. Admittedly, much of that water is running anyway, but it would, under noncotton conditions, improve rather than deteriorate topsoil cover. In high-precipitation areas some 44 kg of topsoil are lost per kilogram of cotton produced.

Material flows relating to cotton production have become an explicit ecological concern to some textile manufacturers, including the Brinkhaus factories in Warendorf, Germany. They welcomed investigations into their life-cycle material flows which were done by Christiane Richard-Elsner of the Wuppertal Institute. The easiest approach was first to study material flows at the plant.

In 1987, Brinkhaus water consumption at the plant level was estimated at 165 l/kg of cotton products sold (mostly bedsheets and apparel). In addition, 2.4 kg of other materials and 6.3 kWh of electrical energy were involved per kilogram of cotton. Since that time, water consumption was reduced by 80 per cent, wastewater by 92 per cent and energy by 13 per cent. If waste heat is used for heating surrounding buildings, another 60 per cent gain is expected in the energy efficiency.

More can be achieved by working on the life cycle of water consumption. By switching from cotton produced in high-precipitation countries to cotton from semiarid areas where subsurface drip irrigation is in use, topsoil losses can be reduced by more than a factor of 15. Another possibility, not yet systematically explored, is soil-protecting cultivation in high-precipitation areas.

2.9 Reducing Material Flows in Industry

The stories of industries that have greatly reduced, even eliminated, flows of various wasted materials are legion. Many industrial managers now understand that waste is simply a resource out of place – a symptom of bad management that hurts the bottom line. Eliminating waste and turning unwanted, often hazardous, by-products into valuable coproducts is now a common and lucrative way to boost profits: in the US, for example, 'a 1992 cross-industry survey of 75 case-studies of pollution prevention found an average payback for industrial investments in waste reduction of only 1.58 years: an annual return on investment of 63 per cent' (Romm, 1994, p 4). Some firms' accomplishments – 3M, Dow, Du Pont, and many more – are almost legendary. For example, Xerox Corporation is on track to achieve a 90+ per cent reduction in waste by 1997 (Romm, 1994, p 145), and AT&T has already reduced its toxic air emissions by 95 per cent and its CFC emissions by 98 per cent.

Here we focus on a few examples of waste reduction and pollution prevention that have the specific effect of reducing the flows of materials in a wide variety of industries:

Harrah's Hotel and Casino in Las Vegas, Nevada, started asking its guests if they wanted their sheets and towels changed every day. Unexpectedly, 95 per cent said they were glad to have been asked, and the vast majority of those said no. Energy and water costs for laundering those 1,800 sets of linens a day – not easy in the middle of a blazing desert – fell by $70,000 a year

(Romm, 1994, pp 3–5). Sheets lasted longer. Pollution declined.

'By 1992, Baxter Healthcare was recycling 99.9 per cent of its plastic scrap. Recycling has saved the company $9 million over the last decade' (Romm, 1994, pp 159–160). The company was on track to virtually eliminate its landfilling during 1991–95, had eliminated both CFCs and its prior production of nearly 12,000 lb of waste oil per year (saving $10,000 a year) and altogether earned a $1.7-million annual net profit on its recycling effort.

At Republic Steel's main plant in Canton, Ohio, worker-driven improvements cut water use by 80 per cent in 2 years, saving about $50,000 a year (Romm, 1994, pp 52–53). The savings ranged from the small (replacing two drinking fountains with water coolers) to the larger (fixing leaky pipes) to the enormous (recycling rinsing water rather than using it once).

At a Ciba-Geigy plant in New Jersey, two improvements in a dye-making process boosted yield 40 per cent, reduced iron waste by 100 per cent, reduced the process's total organic carbon waste by 80 per cent and saved $740,000 a year (Romm, 1994, p 131). An additional 15 per cent yield improvement was later identified.

Even better than making dyes more efficiently is not needing them at all. Natural Cotton Colours, Inc of Wickenburg, Arizona, has built a $5-million business in less than 5 years selling a newly developed strain of cotton that grows in colours and has the same fibre quality as most white cotton. Some colours of this new cotton, called Foxfibres, are cheaper than dyed white cotton and are grown organically. While eliminating the costly, polluting and energy-intensive process of textile dyeing, Foxfibres also have unique properties not found with other cottons: they are fire-resistant, and their colour intensifies with washing. Companies such as L L Bean, Esprit and Seventh Generation are just a few of the well-known clothing manufacturers now using Foxfibres in their products.

Haworth, Inc., a furniture maker, once used 30 gallons of organic solvents per day for cleaning, costing $30,000 a year to buy and a further $9,000 a year to dispose of. Installing two simple batch stills recovered high-purity solvents, reducing the use of one fourfold (with a 1-year payback) and the other tenfold (with a 1.5-year payback) (Romm, 1994, pp 159).

Hallmark, a major greeting-card company, has reduced its emissions of volatile organic compounds from its very extensive printing operations by more than 80 per cent since 1980, mainly by switching to water-based inks. Since 1990, the company has reduced its solid waste generation by 62 per cent, heading for a 70 per cent goal by the end of 1995. In collaboration with Rocky Mountain

Institute, Hallmark is also saving half of lighting energy in new construction and retrofits.

At a Coors refillable bottle-washing facility in Golden, Colorado, the use of citric acid to neutralise caustic label-removing solutions has been eliminated. Workers at the facility and at a nearby Coors can plant suggested that the can plant's sulphuric acid wastes could be used to replace the citric acid. Implementation cost Coors $2,000 and saves $200,000 in citric acid purchases alone, while reducing worker exposure to acids. Similar efforts at a Coors packaging plant eliminated the need for phosphoric acid through better materials management, costing zero to implement and saving $12,000 a year.

Coors has also reduced its manufacturing operations' generation of hazardous wastes by over 75 per cent since 1987, despite increased production. The same reduction was achieved for silver and mercury in wastewater. The large reductions enabled Coors to shut down its licensed hazardous waste management facility; the company expected all of its facilities to achieve 'small quantity generator' status in 1995. Coors has also reduced its releases of potentially harmful chemicals which are included in federal Toxic Release Inventory reporting requirements by over 57 per cent since 1989.

Kleen Test at Port Washington, Wisconsin, the manufacturer of moist cleaning towels, was contacted by the DowBrands Contract Operations team from Mauldin, South Carolina, for a waste reduction programme especially for polyvinyl chloride (PVC) and polypropylene waste. Production line waste has been reduced by 77 per cent and is on a strong continuous improvement track. Moreover, 85 per cent of Kleen Test waste is recycled, and the rest incinerated. Recycling in this case means actually selling much of the plastic scrap to local companies that make a variety of useful products from it, including fence posts and paint roller trays. The project was part of the Dow Chemical Company's WRAP (waste reduction always pays) programme.

2.10 The FRIA Cooling Chamber

Ursula Tischner comes from the design school of Wuppertal University. She earned her degree by designing the functions and the hardware for a new cooling concept that was inspired by Friedrich Schmidt-Bleek's MIPS concept. How can energy and material consumption be drastically reduced without compromising on the cooling function in the household? This was the guiding

Figure 10 FRIA, the cooling chamber designed by Ursula Tischner.

question for her dissertation.

The result has been stunning. She calls it FRIA. Conceptually it derives its location within the house from a larder, or root cellar. The larder has always been a fixed, nonmoving element in the architecture of a house, well insulated against cooking and heating stoves and typically facing north in the Northern Hemisphere. In countries like Germany, the larder is as cool as a typical refrigerator for 3 to 5 months of the year.

FRIA is in a sense a multichamber refrigerator with certain design principles borrowed from the larder concept. But FRIA uses high-tech cooling techniques and more and better insulation than conventional refrigerators. It has an extremely long service life, and the chambers can be repaired or exchanged separately. Deep-freezing boxes can be included. Figure 10 shows the design of a standard model of FRIA. It is convenient, functional and appealing.

Depending on the cooling device used in FRIA, the energy savings can also be considerable. The standard model FRIA in 1994 would use no more than 0.40 kWh in 24 hours as compared with 0.85 kWh for more conventional German refrigerators. If the cooling unit installed in FRIA were made by Gram (see p 34), energy consumption could go down to 0.26 kWh/day. A third option would be to integrate the Zeolith hot-water heating system into FRIA. This system, invented by Zeotech, a Munich-based firm, supplies hot water with 30 per cent better efficiency rate than the industry standard for hot-water heaters, with the additional benefit that substantial cooling capacity would come as a welcome by-product.

Because FRIA can outlive perhaps five to ten generations of traditional refrigerators, as well as being at least two to four times more energy-efficient, a factor between four and eight in total resource efficiency compared with conventional refrigerators is achievable.

2.11 The Services of Washing and of Vertical Transport in Buildings*

Many goods and installations in buildings can be optimised for low resource use. By using the right incentive system to shift from large numbers of short-lived goods to a small number of more durable goods, the services provided can be dramatically de-materialised. Let us illustrate the potential in buildings by two examples: washing machines and lifts.

Laundromats versus Individual Washing Machines

Modern multi-apartment buildings in Northern Europe, as well as in the US east and west coast states, tend to have laundromats to discourage families from using washing machines in their apartments (Stahel and Gomringer, 1993). The average gain in energy efficiency per washing service rendered has been estimated to be about a factor of four. The material productivity gains are even larger, probably in the vicinity of a factor of ten. The resource-saving strategy is to sell shared utilisation of few machines rather than selling many privately owned machines.

The energy gains come mainly from the choice of the energy source. Privately owned machines almost always run on electricity. Often, electricity-driven tumble dryers also add to the power bill. Laundromats, on the other hand, tend to use natural gas to heat the water (a far more efficient and cost-effective method) and can partly reuse the hot water or recover the heat, thanks to a higher washing cycle frequency. Also, the waste heat is then available for reuse in tumble dryers.

Gains in material productivity stem from the more intensive use of the laundromat. Built for robustness, the laundromat machine typically lasts for 30,000 washing cycles. Household machines with similar life-cycle materials consumption last for only an average of 23,000 washing cycles.

---

* We owe the substance of this subchapter to Walter Stahel.

The social drawback of the laundromat is its positioning in unattractive places such as chilly and lonely places in the basement or unpleasant rooftop locations. In California, a new generation of laundromats has solved the problem. In condominium complexes, they are located next to the swimming pool and barbecue areas.

Many other household functions can be dematerialised by manufacturing more durable machines, by arranging shared use or by principally switching from a goods market to a service market as described by Giarini and Stahel (1993).

LIFTS

To illustrate the service principle further, let us look at 'vertical transport', *i.e.* lifts. Tall buildings (and hence space-saving construction) are inconceivable without efficient lift systems. Lifts chiefly consist of a set of rails, a cabin running on these rails, a counterweight, a traction engine with a gearbox, a control mechanism and an emergency-braking system. Due to their modular construction and the fact that most of the technology is hidden, lifts are long-lived goods which can be comparatively easily upgraded to new technologies (such as electronic controls and thyristor-controlled motors) and fashion (changing the service panel, the decoration or ultimately the entire interior of the cabin) (Stahel and Gomringer, 1993).

The lift company Schindler AG in Ebikon, Switzerland, number two in lifts worldwide, wanted to make use of the theoretical longevity of its lifts. Schindler therefore began to sell 'vertical transport' as its product, *i.e.* to lease its lifts. The leasing contracts include regular maintenance and services, thus guaranteeing to clients the avoidance of the frustration of a stalled lift. Clearly, the commercial interest of the firm is now derived from its lifts' excellent longevity and reliability. By 1992, 70 per cent of Schindler's revenues were coming from such service activities.

By their systemic durability, lifts can be said to be roughly 40 times more energy-efficient and some 10 times more materials-efficient than an average 1995 automobile.

2.12 REHABILITATING VERSUS DEMOLISHING BUILDINGS

Old buildings are often useless for a new proprietor, or no longer economic because the zoning laws may have changed since their construction. The result is often that buildings are pulled down and new ones are constructed in their place. With no loss of service, a factor of four in both energy and materials could typically be saved by modernising the old building. Also, in the latter case, societal and emotional values such as cultural heritage and the atmosphere of a familiar ensemble of buildings can be preserved.

The factor of four results mostly from the preservation of 'grey energy' contained in the building's load-bearing structure, *i.e.* in bricks and mortar. Even if all technical installations for heating, cooling, power, lights, lifts and windows are replaced by modern substitutes (hopefully more efficient – see pp 10–29), 75 per cent of the energy and materials originally invested will be preserved.

Labour, on the other hand, will not be saved: quite the contrary. Renovations can be just more labour-intensive than pulling the old building down and making a new one from scratch. The cost balance will shift significantly towards renovation, and thus resource efficiency, if the tax burdens on human labour are reduced while resource taxes are introduced (see Chapter 7).

Reuse of Materials from Demolished Buildings

When a building is demolished rather than renovated, what happens to its materials? Most, like construction wastes, go into landfills that are rapidly filling up. For example, the Greater Vancouver Regional District in British Columbia, Canada, produces 1.4 million tonnes per year of general municipal waste. Construction and demolition wastes add, in roughly equal proportions, another 0.3–0.6 million tonnes a year, most of which are landfilled. With 60 per cent of B.C. landfills expected to close by 2000, the Ministry of Environment has called for at least a 50 per cent reduction in municipal solid waste by then. Tipping charges accordingly doubled from 1989 to 1992, with a further more-than-doubling planned by the end of 1995 – similar to the fivefold increase experienced in Toronto over a recent 5-year period.

Such dramatic increases in disposal costs concentrate builders' minds wonderfully. One result was a pilot project by British Columbia Buildings Corporation – an outstanding Crown Corporation that develops and runs buildings, chiefly for the public sector – to try out a new concept: environmentally responsible demolition. The building tested was the 1963 Westgate Annex building of the old Oakalla prison complex. This 24-by-46 m building had a concrete floor, concrete-block walls, wood joists and roof decking, interior wood-and-drywall finish and barred windows.

After asbestos removal, the pilot project was launched by a request for proposal to interested and capable contractors, with the express aim to 'channel waste away from landfills, toward reuse and recycling, while ensuring that the project was economically compatible with the normal method of building demolition.' Bids were to specify for each type of material the quantity, disposal price, alternative waste management price and companies that would accept the materials for reuse and recycling, and to include two prices: one for a normal demolition and another that maximised opportunities to recover materials. The result: a winning bid that showed the recovery-reuse-and-recycling option was 24 per cent cheaper than normal demolition.

The 1991 dismantling was highly successful, using labour-intensive methods where appropriate. Three-quarters of the concrete wall blocks were reused by a boys' club for new structures, the rest crushed for aggregate. The wood beams, decking and other lumber were 97 per cent recovered and resold, as were the metal, miscellaneous equipment, drywall (sent to a gypsum plant for recycling), windows, bars and other valuable items. Loose gravel was moved to one end of the roof with a fire hose and taken away by a landscaper for a walkway. Footings rich in wire, and hence unsuited for recycling as aggregate, were sent to a fill site for reuse as a road sub-base. The only significant material landfilled was 61 cu m of friable roofing (the glass fibres of which had been impregnated with hot tar, making intact recovery impossible) and 46 cu m of wood debris too broken to reuse (although in some communities it could still be valuable, at least for fuel).

Of the total volume from demolition, 64 per cent was estimated to be wood, 30 per cent concrete, 2 per cent metal, and 3 per cent the tar-and-gravel roof. Normal demolition would have sent 92 per cent of this entire volume to landfill. But in the pilot project, only 5 per cent was landfilled and the other 95 per cent reused or recycled. The 1.5 months of extra labour for the demolition crew was compensated for by the sale of the materials. The contractor believes that, although this building contained an unusual amount of valuable wood, the materials-recovering approach would probably still be profitable even with other buildings without so much wood, wherever there is a market for the materials. It is only necessary, he concluded, to set the goal, plan its implementation and leave extra time.

2.13 Perennial Polyculture

Reinventing Agriculture

Agriculture bothers Wes Jackson. Not how we do agriculture, but agriculture itself. At the risk of being accused of wanting to return to a Stone Age life, Jackson questions the impact and sustainability of agriculture as we know it, that is, tillage agriculture. Ripping up and turning over the soil has profoundly changed the earth. 'Geologically speaking,' he says, tillage agriculture 'surely stands as the most significant and explosive event to appear on the face of the earth, changing the earth even faster than did the origin of life' (Jackson, 1980, p 3).

So profound is the ecological disruption of tillage, yet so familiar to Western concepts of what agriculture is, that it is disturbing to hear Jackson tell of the Native American who watched as a 'sodbuster' farmer, newly arrived on the Great Plains of the Midwest, drove his team of horses to plough up the virgin prairie. Without expression, the Indian stared at the mouldboard plough, slicing through the thick mat of prairie grasses and turning them under, roots in the air. After a long time, the farmer paused and asked the Indian, 'Well, what do you think?' Replied the Indian, 'Wrong side up' – and went away.

It's still wrong side up. In a few thousand years, tillage agriculture has transformed vast areas of the earth from rich spatial symphonies of diverse plant life to patchworks of monocultures. What's more, soil losses that accompany tillage – in some places rapid, in others slow but inexorable – cannot be indefinitely sustained. In recent years a dumper truck-load of topsoil per second has passed New Orleans in the Mississippi River. At the rate we're going, western Iowa will be out of topsoil (already half gone) before western Nebraska is out of groundwater. What agriculture needs, says Jackson, is a 'bio-technical fix.'

Perennial polyculture is Wes Jackson's goal. At the Land Institute in Salinas, Kansas, Dr Jackson, a plant geneticist, gathers forward-looking plant researchers. Together, they are learning from the American prairie – a diverse ecosystem of hundreds of kinds of perennial plants where soil 'stays put'. Plant-breeding experiments at the Land Institute seek to develop perennial grain crops to replace the annual grain crops that are the basis of tillage agriculture. Perennial crops virtually eliminate the need for soil-eroding tillage. Growing them in a polyculture will encourage diverse soil flora, fauna and microbes essential to natural processes of organic decay that sustain soil fertility.

The textbooks say that perennials cannot be high yielders. The textbooks are wrong. It took Land Institute researchers only a couple of years to show, and a decade to prove with full rigour, that even 'eyeball-selected' cultivars from ordinary backyard prairie grasses could match or beat the seed and protein yields per acre of the highly hybridised, genetically vulnerable, input-intensive cereals that have taken a century to breed, yet cannot survive weather and pests without constant attention. Thus was vindicated the Land Institute principle that 'if there were a better way to use the sunlight than the prairie that grows here, it would have been here already.' Using nature as model and mentor, not as a nuisance to be evaded, yields the rich dividends of respecting several thousand million years of design experience in which everything that didn't work reliably was recalled by the Manufacturer, and most living things today, though works in progress, are already well 'debugged' by nature's free testing and design-improvement service.

Not content with the herbaceous seed-bearing perennials that stretch to the Kansas horizons (in the few places where native prairie survives), the Land Institute team started doing wide crosses. Eastern gama grass, Illinois bundleflower, Siberian giant wild rye – all yielded astonishingly hardy, versatile and high-yielding crops with relatively little effort. Like their native forebears, the seeds of these new crops could make delicious bread, either mixed or separated by conventional mechanical techniques. Grown in combinations of several plants together, they could look after each other, with one fixing nitrogen, another exuding protective herbicides, yet another guarding against insect attack. Of course, any polyculture is less attractive to the infestations that plague monocultural crop systems.

The ultimate result of this nature-imitating vision of agriculture will be fertile fields that look rather like a prairie. The crops, of many different kinds all mixed together, come up every year with no tillage, no sowing, no soil erosion. They grow with no irrigation, no fertilisers, no pesticides. When they're ready, they're harvested, either mechanically by a grain combine or by the native ungulates like bison and antelope that evolved optimally to eat such grasses.

A biochemical firm expert to whom this system was recently described stood nonplussed:

'When do you spray it?'
'You don't.'
'Fertiliser requirements?'
'None.'

'Well, what do you *do* to produce it?'
'Nothing. You sit there and watch it grow.'

Factor 10, factor 100 – how big are the ultimately available resource savings in water and energy and agrochemicals? Nearly infinite, because other than a trivial amount of harvesting energy, and the little energy it takes the farmer to walk or ride around to maintain a healthy eyes-to-acres ratio, there are no inputs.

It may take another half-century or more to develop something like the perennial polyculture of human food crops Jackson envisions. But all the basic science needed to show its feasibility has already been demonstrated. What remains is a few generations of details.

Jackson is looking ahead, but he is also concerned with the near term. 'Long before we get to the point of fine-tuning such a system, our farmers will need to employ the entire array of sound soil conservation measures as a holding action' (Jackson, 1980, p 131). To this end, the Land Institute is also researching innovative practices in conventional agricultural systems. The Sunshine Farm project is one such endeavour. Using low-input principles, the best tillage practices, photovoltaics and wind turbines, the Institute's Sunshine Farm researchers are moving towards farm energy self-sufficiency. One question: What amount of oil seed production is necessary to fuel the tractors needed by the farm operation so that it yields net food while consuming no fossil hydrocarbons?*

Over time, the Sunshine Farm project will be integrated with the perennial polyculture research. Jackson and his colleagues believe possible no less than replacing an energy-consuming and soil-losing agriculture with an energy-producing and soil-building one. On this Second Agricultural Revolution, the one that stops the war against the earth and reopens diplomatic relations between nature's wisdom and human cleverness, may turn the ability of our descendants to eat well, or at all.

2.14 Biointensive Minifarming

A long trend of ever-greater gains in global agricultural productivity has recently faltered; we have entered an era of decreasing rates of increase in yields and even decreasing absolute yields (*State of*

* The Land Institute *Annual Report* 1993, Salinas, KS 67401 (USA), tel 913-823-5376.

the World 1994 and *Vital Signs 1994*, Earthscan Publications, London, 1994). Soil erosion, decreasing soil fertility and diminishing returns to greater off-farm inputs all contribute to the productivity problem and the unsustainability of conventional agriculture. Yet solutions may come from unlikely places.

On a steep hillside in northern California, on soils many would consider unfarmable, John Jeavons and his colleagues at Ecology Action are developing a system for highly productive and sustainable agriculture (Jeavons, 1991; Jeavons and Cox, 1993). Jeavons and colleagues are pioneering what Jeavons calls 'biointensive minifarming'. Their system rests on four main principles: deep cultivation of the soil to provide for optimal root growth; production of compost crops to feed back to the soil; intensive spacing of plants in wide beds to create beneficial microclimates; and interplanting of different crops to foil pests. The system is not mechanised, yet its labour requirements are surprisingly small. While the biointensive beds take some effort to establish, once the elements of the system are in place, maintenance is simple, because nature does most of the work.

The Ecology Action researchers are not simply interested in growing more vegetables. Their goal is to develop the techniques needed to produce a person's entire calorific and nutritional requirements sustainably and in as small a space as possible. To this end, they are experimenting with intensive production of grains, legumes and other high-calorie crops. Much of their effort is also directed towards compost crops – essential for building rather than depleting soil.

According to Ecology Action, mechanised and input-intensive agricultural practices in the US require 45,000 sq ft or more to feed a person on a high-meat diet, or roughly 10,000 sq ft for a vegetarian diet. Yet in developing nations, by the year 2000 only 9,000 sq ft of arable land will be available per capita. Far less land will be available as population further increases and desertification, soil erosion, urbanization, and other problems shrink the arable land base. Fortunately, biointensive techniques can produce a vegetarian person's entire nutritional needs, plus the compost crops necessary for sustainable production, on only 2,000 to 4,000 square feet of land. These same techniques, compared with conventional agricultural practices, can also reduce water consumption per unit of production by up to 88 per cent, reduce off-farm energy inputs by up to 99 per cent, and double net farm income per unit area.* Other than the land and a few hand tools, initial capital is virtually nil, and no chemical inputs are required.

* *Annual Report* 1993, Ecology Action, 5798 Ridgewood Road, Willits CA 95490-9730 (USA), tel 707-459-0150, fax 707-459-5409.

Many of the techniques for biointensive, miniaturised agricultural production have been known in China and other parts of the world for thousands of years. Some East Asian systems that combine agriculture with aquaculture obtain even higher, almost unbelievably high, yields of food calories and of protein from very small areas. Typically they stack up several systems vertically: for example, rabbits whose wastes fall into and fertilise a fish- and duckpond, which in turn is coupled to rice and vegetable beds, whose wastes feed the rabbits. Other schemes involve two rice paddies which are alternately flooded and drained, alternating rice with fish, shellfish and ducks.

Ecology Action has been experimenting with these practices, carefully documenting successes and failures, inputs and yields, since the early 1970s. In recent years, Ecology Action has established an aggressive outreach programme to teach the biointensive minifarm system. Jeavons and his colleagues take their accumulated experience on the road, giving workshops across the world. Others have joined their efforts. Demonstration centres are now established in Mexico, India, the Philippines, Russia and Kenya, and many more affiliated farms and research projects report their results back to Ecology Action.

2.15 Rent-a-Chemical: A Strategy for Improving Material Efficiency*

Chlorinated Solvents

Chlorinated hydrocarbon (CHC) solvents are extremely useful. They truly contribute to the quality of modern life. We use them as cleaning (degreasing) agents, as adhesives, for chemical extraction and as solvents in the textile, pharmaceutical, plastic and metalworking industries. They are chemically quite stable, they don't burn and they are not soluble in water – all rather convenient properties for the purposes for which they are used.

Some 1.2 million tonnes are produced annually, which is why their use has attracted so much attention from environmentalists. Precisely the same qualities which make the chemicals so useful in industrial applications also render them hazardous to human health and the environment. Their excellent grease dissolution ability

* We owe the substance of this chapter to Sascha Kranendonk of the Wuppertal Institute, who received her information mostly from Dow Europe.

combined with their chemical stability allows them to enter the fat tissues of humans and animals. CHC solvents are proven to be liver-toxic and are suspected to be carcinogenic. Of the 600,000 tonnes of chlorinated solvents sold in Europe in 1992, only about 90,000 tonnes, 15 per cent of the sales, were recycled. Some 450,000 tonnes evaporated and thus contributed to air pollution, some 20,000 tonnes were incinerated (hopefully in modern incinerators without dioxin emissions) and some 40,000 'disappeared' – unaccounted for, probably to the detriment of the groundwater in many places. Groundwater clean-up after CHC spills is an extremely tedious and costly business.

All these environmental troubles have led German lawmakers to adopt regulations to limit evaporation and to make take-back obligatory for chemical manufacturers. An unpleasant result of the CHC phobia for the manufacturers was a steep decline in CHC sales.

One way out for manufacturers, which at the same time can be very beneficial for the environment, has been a new marketing philosophy called 'rent-a-chemical'. The idea was introduced by Dow Germany through SafeChem, a joint venture with a local recycling company, RCN. The idea is that the chemicals manufacturer, Dow, keeps control over hazardous chemicals throughout their life cycles. It is also a way of applying the total product responsibility principle (Fussler and James, 1996).

SafeChem stores and transports the solvents in specially designed ultrasafe containers, and an air-tight pumping system is used to avoid evaporation (which is no longer allowed under tightened German air pollution legislation). Direct risks to the health of the handlers are also eliminated by this method. Taking control of how the chemicals are used and handled enables clients to regain a percentage of the used solvents, but, in addition, it is possible to reuse and recycle them many more times than would otherwise be the case. SafeChem provides refill stabiliser and a solvent-test tool kit to maintain the quality of its product as long as possible and at every stage.

This way, solvents can be recovered and reused more than a hundred times. The material efficiency of the packaging would not be quite that much due to the material inputs that are unavoidable for manufacturing the safe-handling system, but a factor of ten is easily attained. And, after all, the ecological motif has not been materials efficiency but pollution prevention, or keeping the convenience of extremely useful chemicals without incurring hazardous (and now illegal) air and water pollution.

When SafeChem first introduced the system, both clients and competitors remained sceptical. But the system worked well. And, as of early 1995, competitors have been hastening to follow the leader. Dow is already looking into other chemicals that lend themselves to the leasing idea.

There are plans to go one step further, quite along the lines of the Product Life Institute, by offering a service instead of a solvent. The service could, for example, be termed 'degreasing per square inch'. The commercial interest in producing solvents would thereby further decrease, and the interest in avoiding losses and pleasing both clients and the environment would be higher still.

2.16 Using Less Concrete for Stabilising Walls

Imagine you are planning to build a new home very close to or abutting your neighbour's home. Such is the typical situation in densely populated European cities. To use the space as efficiently as possible, you will probably want to have a basement, so the construction firm will dig a deep hole in the ground. If nothing is done to stabilise the facing wall of your neighbour's house, it is likely to weaken and may even come down in the process of your excavation and construction.

Several techniques are available for stabilising the neighbour's wall. Essentially two types of 'underpinning' (shoring up) are in use: underwall and facial (*i.e.* in front of the wall). Professor C. J. Diederichs of the Technological University of Wuppertal asked two students, F.J. Follmann and T. Schröder, to investigate the difference between the two methods in terms of material consumption. They used the MIPS calculation of Schmidt-Bleek (Chapter 9) and found that the more conventional underwall techniques (high-pressure injection or a patent pile system) involved roughly four times as much material flows as the more advanced facial techniques – 'soil nailing', profiled walls such as tongue-and-groove and drilled-piling walls.

The tabulation below presents the results, published by Diederichs and Follmann (1995). For each technique ecological rucksacks were calculated for all materials directly involved in the process, such as process fuels, steel, cement and other additives used in concrete manufacturing. Water and compressed air constitute major contributions to the rucksacks. All data refer to the rucksacks in tonnes per lineal metre of adjoining wall. Energy consumption is also significantly lower for facial shoring-up than for the underwall methods.

Table 2 Five different techniques for stabilising the neighbour's wall. They differ greatly with regard to the material flows they involve.

Method of underpinning	*Mass of the construction*	*Material flows*	*Process water*	*Process air*
High-pressure injection	35.0	44.5	80.4	77.8
Patent pile system	25.2	42.1	63.8	42.9
Soil nailing	3.2	7.7	11.9	8.7
Profiled wall	5.5	10.7	16.8	12.0
Drilled-piling wall	4.4	7.4	11.6	9.0

The authors are honest about the total ecological impacts. Because the underpinning is only a small part of the construction, the savings for the whole new building remain rather modest. Even the most materials-efficient facial underpinning requires preparatory activities whose total impact may exceed that of the underpinning itself. The least impact might result from doing the entire building without a below-grade level, as is the case in most Dutch buildings – requiring, however, higher buildings or larger land area per useful square metre.

Even if the underpinning remains a small portion of the construction process, it can be assumed that other elements of the process are also susceptible to improvements in material and energy efficiency. It has been estimated that the energy intensity of modern construction is roughly a hundred times higher than that of preindustrial construction.

2.17 BELLAND MATERIAL: RECYCLING WRAPPING PLASTIC

Plastics have been a nightmare for the waste disposal business. Plastics typically don't rot, and they look so ugly on landfills that they invariably have been the targets of the accusing cameras of environmentalists. But burning them has not been that much more attractive. Flames can convert chlorine and other halogens that are often contained in plastics into dioxins and other highly toxic substances. Very high temperature furnaces can destroy dioxins, and modern scrubbers can clean the exhaust gases. But all this is not without failure, it's expensive and it's very unsatisfactory from the point of view of resource efficiency.

So communities are turning to plastics recycling. But mixed plastic waste leads to deterioration of properties ('downcycling'). Pure plastic is difficult to obtain from household waste. Mechanochemical separation of shredded mixed plastic is possible to a degree but not 100 per cent, and it's very expensive. Certain American states, *e.g.* Vermont, let their residents separate household plastic waste into up to seven different waste containers. But can that be a solution worldwide? What percentage of households would have enough space to accommodate seven dustbins for plastic and three more for organic waste, paper and metals or glass?

The German approach makes for easier handling on the part of the household. All packaging materials, whether metal, plastic or composite blister wrapping, are supposed to bear the Green Dot, certifying that the recycling fee was paid beforehand. All Green Dot materials go into yellow bins. For wastepaper blue bins were introduced, and the remaining waste goes into smaller black bins. (Glass is expected to be brought to 'igloos' that are publicly accessible in the main shopping areas.)

But the plastic collected in the yellow bins has caused no end of troubles. It was shipped abroad, sometimes as far as Indonesia. It was burnt (against the rules of the Green Dot) or landfilled (also against the rules), or downcycled into noise-protection walls (not an unlimited dumping facility). Finally, expensive installations were established for chemical decomposition of plastic down to 'recycled crude oil', which then would be allowed to burn. All this has made the Green Dot system a laughing stock in Germany. And foreign manufacturers wanting to export goods packaged for end consumers in Germany find the system a massive nontariff trade barrier and a nuisance.

So what can we do? Do without plastics, as some green fundamentalists suggest? Certainly not! Modern supermarkets require hygienic wrapping of all items and full visual transparency for customers. Many would maintain that given such conditions there is no realistic alternative to plastic and blister packaging.

But there is an alternative to traditional polyvinyl chloride (PVC), polyethylene (PE) or other wrapping plastics: Belland material, developed by Roland Belz, a German engineer living and working in Switzerland. Belland plastic has a very convenient property: at pH values a bit higher than seven, it becomes water-soluble. Virtually all other typical properties of plastics can be engineered and obtained with Belland plastics, such as transparency, elasticity and different degrees of stiffness for uses in soft wrapping foils, up to mechanically robust parts.

In a society using Belland plastic in a variety of functions, including the wrapping of goods for final consumers, the material would be found in smaller or larger quantities in all dustbins. By rinsing the collected contents with basic water, all Belland material can be recovered. It will all be contained in the effluent water. By adding a few drops of citric acid or any other harmless low-pH substance, the Belland material can be made to coagulate. The precipitated material can easily be collected and turned into chemically pure granules for further processing.

In a country or municipality managing to maintain a separate collection of wastepaper, Belland plastic waste should go with that paper. In a simple first step that is routine in the processing of wastepaper to pulp, Belland plastic can be separated at minimal added cost. The breakthrough for Belland might come with dishes and cutlery for fast food and catering at big sporting events or trade fairs. Dishes can be collected after use in their typical messy state and can be conveniently processed for remanufacturing, leaving no trace of the organic waste. At the world's largest international plastics fair 'K', Düsseldorf, November 1995, the Belland catering system was successfully tested.

As is the case with aluminium, the recycled material has precisely the same properties as the virgin material, but far less energy and matter are needed to produce the recycled version. On a life-cycle basis, a factor of four to ten is easily achieved in material efficiency when changing from, say, PVC to Belland material.

2.18 REUSING BOTTLES, CANS AND LARGE CONTAINERS

Even better than recycling the material for containers is not to destroy the containers in the first place. Refund–reuse systems are in wide use for mineral water and beer bottles throughout Northern Europe. Statistically, bottles are used some 20 times, in some cases 50 times, before being destroyed by accident or on purpose (because scratches may have rendered them unappealing or because the cap is no longer airtight). Compared with disposable glass bottles, aluminium cans or compound plastic containers, a factor of four can easily be reached in total resource efficiency. However, there are two major exceptions to this rule. Milk and other protein-containing packaging requires very intensive cleaning before reuse in order to prevent dangerous microbiological contamination. Heating water or steam for the cleaning and using detergents can easily outweigh the ecological benefits from the reuse of glass containers.

The other exception relates to distance. Shipping empty glass bottles over distances above 250 km or so is not ecologically sound. Hence, refund systems should preferably be regional. Highly centralised beverages and food industries don't fit well with refund systems.

But waste activists need not give up on resource-efficient reusable containers. The Munich-based waste activists of Das Bessere Müllkonzept ('the better waste concept') are confronting food centralisation head-on. They demand regionalised food supplies to the degree possible and suggest standardisation of all food and beverage containers. They urge consumers to use baskets and household containers for daily shopping wherever possible, and encourage local farmers and food processors to offer fresh produce rather than canned food. But if long-term preservation of food is needed, the cans or other containers should without exception be returnable on a deposit–refund basis. The waste activists claim that such a system is technically feasible, that cans can be systematically collected, washed and reused observing all hygienic requirements of local suppliers. Clearly, one can anticipate a coalition forming between waste-watching groups and local organic farmers whose income depends on 'green' customers, preferably close enough for profitable direct marketing.

There is always an element of protectionism involved in schemes requiring the use of returnable containers. When in 1987 Denmark introduced the obligatory use of returnable bottles for a wide range of beverages, the European Commission, after lobbying by French producers of mineral water, sued Denmark for violating the rules of free trade in the Community. But the European Court in Luxembourg essentially ruled in favour of Denmark, thus giving an important precedent to national laws protecting the environment even at the partial expense of free trade.

REUSABLE CONTAINERS FOR CAR PARTS

Henry Ford's original car factories had a whole section devoted to reusing wooden crates and pallets. But now the auto industry, which ships enormous volumes of parts all over the world, has an even better idea: reusable steel cases.

In April 1994, Mitsubishi Motors Corporation – the first Japanese automaker to establish an in-house Global Environmental Issues Project Team to lead a 'green' transition – collaborated with its German distributor, MMC Auto Deutschland GmbH, to use steel instead of wooden and cardboard packing cases. MMC ships about 2,800 cases of replacement auto parts per month from Nagoya, Mizushima and Takatsuki to Germany. After a month-long sea voyage, the cases are unpacked at MADG. Previously, the cardboard and wooden packing materials were thrown away. Now the new steel cases are simply emptied, folded down, sent back to Japan, and reused – for an expected 10 years.

A similar switch to stackable, instead of single-use, packing crates that are returned and reused is expected in the next few years to reach a 95 per cent reusable container rate at Diamond-Star Motors. This firm also reuses case bottoms, which are shipped back to Japan for a further three to four trips, and specifies in its Suppliers Packaging Guide that its more than 500 North American suppliers are to use reusable packing materials and bins whenever possible.

2.19 WIDE-SPAN, HEAVY-DUTY WOOD CONSTRUCTION

Wood is an astonishing construction material. It is light, attractive and natural. Properly selected, treated and maintained, it is more reliable and more durable than concrete in the most common construction functions. Its production uses less than one-quarter (Factor Four!) of the energy required by concrete. For all practical purposes, it is renewable if sustainably grown and harvested; and certification systems are emerging to enable buyers to ensure that the wood they use has these important qualities.

The usually cited drawbacks are that wood is not as suitable as other materials for wide-span heavy-duty functions, that it's old-fashioned and that forests would suffer steady timber extraction if wood were again to be widely used for construction purposes.

'Wrong, wrong and wrong,' says Julius Natterer, a Bavarian who teaches wood construction technology at the Federal Institute of Technology in Lausanne, Switzerland. 'These drawbacks don't exist.' Natterer does impressive wide-span wood constructions, of which Plate 8 gives an example.

But Julius Natterer doesn't restrict himself to showpiece projects. He also offers down-to-earth affordable wood construction for apartment buildings. Their construction shows an excellent energy balance compared with concrete and brick buildings. And all the energy efficiency requirements specified in Chapter 1, pp 10–29 can be fulfilled.

If wood and masonry buildings are compared for their inputs of nonrenewable mineral resources, the balance in favour of wood evidently becomes much better still. A factor of ten is doubtless available.

What about the timber extraction from forests if there were a massive shift towards wood use in the building industry? Natterer has thought about this question and, for an answer, he points to the positions taken by Swiss politicians who are known for their strictness about environmental protection. Says Pierre Aguet, national deputy from the canton of Vaud: 'We could sustainably harvest seven or eight million cubic metres of timber from Swiss forests, and it would be a blessing for those forests. It would finance all the necessary forest protection measures. Those seven or eight million cubic metres would be sufficient for building some 250,000 Natterer-style apartments annually, vastly more than Switzerland will ever need.' Fifty other national deputies from all political camps have associated themselves with Aguet's position.

Other European countries could do the same. Even tropical timber use is not necessarily a bad thing *per se*. It all depends on truly sustainable forestry practices and on prudent and efficient use of the material.

2.20. Wood in Home Building

Some 90 per cent of American houses are 'stick-built' – a traditional, labour-intensive and in many ways rather primitive method. Outside walls are framed with vertical 'two-by-four' studs – long pieces of wood nominally 2 by 4 inches (5.1x10.2 cm) in cross-section but actually somewhat smaller. To make the wall strong enough, from the centre of one stud to the centre of the next is normally 16 inches (40 cm). That construction should be sound if as little as 10–15 per cent of the opaque area of the wall (*i.e.* excluding window and door openings) is wood.

Actual construction differs. The opaque wall area is normally 30–35 per cent wood – two or three times what is needed. Carpenters are paid by the hour, and may have no incentive to conserve wood they're not paying for. They also have a proverb: 'When in doubt, build it stout.' They add wood everywhere: odd sill and corner details, 'cripples' (diagonal bracing filling empty spaces), triple and quadruple plates and so forth. In addition, a lot of wood is wasted, because standard stud lengths don't divide evenly into the lengths required by construction that is seldom planned to economise on lumber.

On the other hand, as old forests disappear, the studs are no longer of such good quality. Soon after framing, a tenth of the studs may have to be chain-sawed out and replaced because they have warped so badly that the materials to finish the wall won't fit on.

And, because there's so much extra wood, the fibreglass/mineral-wool insulation between the studs covers less area than it's supposed to, while the extra wood, which conducts heat three times better, leaks more heat through its 'thermal bridges' across the insulation. This can easily reduce the actual insulating value of such a studwall by 20–25 per cent, from R–19 to R–14 (k–0.29 to k–0.40).

The Davis house in the ACT^2 experiment (Chapter 1, p 16) pioneered an important innovation by the Davis Energy Group: an 'engineered wall' designed to do more with less. Instead of normal solid softwood like fir, it used an 'engineered wood' or 'oriented strandwood' product made by TrusJoist MacMillan, a large Idaho firm. This product is extruded under heat and pressure from low-grade, low-density, usually small softwood, such as aspen or poplar, into a solid billet about 8 inches (20 cm) thick and several metres wide. In effect, it's 'synthetic hardwood', but with a strength, uniformity, predictability and freedom from knots and flaws almost unknown in hardwood.

Cut into thin, studlike lumber, the engineered wood product is strong enough that 1.25-by-3.25-inch (3-by-9-cm) studs, mounted on 24-inch (61-cm) centres below a continuous 1.25-by–14-inch (3-by-36-cm) header, are stronger than a normal studwall. (In fact, the original design used 48-inch centres but had to be changed because the local building code required the closer spacing, not to make the wall strong enough – it already was – but rather to fit one of the approved methods for attaching stucco to the outside.) The wood/wall ratio then drops from 30–35 per cent to only 9 per cent, a 70–74 per cent saving that far outweighs the engineered wood's higher cost.

Because fewer pieces must be cut and assembled with less nailing than in a normal studwall, saved wood and labour more than pay for doubling the thickness of the insulation between the studs. The insulation is high-quality, foil-faced, fibreglass-reinforced polyisocyanurate foam that fills the space more exactly, provides a vapour barrier, saves labour, and adds rigidity and soundproofing. The thicker insulation plus the reduced thermal bridging increase the actual insulating value by 85 per cent, to a true R–27 (k–0.21). Air leaks are also greatly reduced. The result: a dimensionally stable, tighter, stronger, more durable, faster-to-build wall that uses one-quarter as much wood, insulates almost twice as well and costs $2,000 less per standard American tract house.

That factor of four is not the end of the potential. Some firms are now sandwiching thin layers of carbon or Kevlar fibre between layers of natural or engineered wood. This more than doubles the strength of the wooden element, making the wood smaller, lighter and producible from lower-grade timber. In addition, BellComb, a Minneapolis firm, has developed a cardboardlike (optionally recycled) honeycomb structure prefabricated in a wide variety of precisely cut shapes and sandwiched between inexpensive strandwood sheets. The pieces fit tightly together like a child's miniature house kit. Two unskilled adults can build a cottage-sized structure from this material in 20 minutes and (if it hasn't been glued together) take it down again in 10. It's airtight, fire-resistant and easily superinsulated by adding foam layers into the sandwich. It also saves about 75–85 per cent of the wood, with still further savings in sight.

Chapter 3
Ten Examples of Revolutionising Transport Productivity

We are devoting a separate chapter to transport productivity, because while each transport of goods or people involves both energy and material consumption, not all environmental impacts of transport relate to energy and materials. Habitat destruction (by roads), noise, mass tourism and ever-increasing access to natural resources involve more than resource considerations, and need to be considered separately. Certainly, transport–environment conflicts bear a special importance that will make any efficiency gains in transportation warmly welcomed even if they save resources. Moreover, the description of ways and means to quadruple transport productivity will yield insights about a new civilisation that we will have to develop anyhow for reasons that go beyond the efficiency revolution.

3.1 Videoconferences

Data highways have become one of the most powerful symbols of technological progress. Al Gore's 1992 best seller, *Earth in the Balance*, has helped to build a broad public awareness of the important role that data highways can play in harmonising ecological and prosperity objectives. In this subchapter we are exploring in preliminary quantitative terms the potential of long-distance electronic communication to help multiply resource productivity.

Not surprisingly, we have found that the potential is far bigger than a factor of four. We did a rough estimate for two different cases:

- substituting electronic mail for posted letters (see p 116); and
- substituting a videoconference for a business meeting (see Plate 9).

Rocky Mountain Institute has been one of many early adopters of the systematic use of long-distance data communications. At RMI, videoconference apparatus that digitally compresses full-duplex colour and audio so it can be sent over a data-quality telephone line joins with e-mail, modem text and graphics transmissions, telephone and fax to displace much travel. Soon after its installation in 1993, for example, the videoconference equipment enabled one of us to avoid 4 days' travel to and from Western Australia, at a few per cent of the cost of the plane ticket, and with none of the inconvenience or fatigue. A major conference could be keynoted, complete with projected graphics, simply by sitting in a comfortable chair at RMI in front of the videocamera, pressing a few buttons to dial the apparatus at the other end and speaking normally.

Microchips programmed with Israeli data-compression algorithms sent images of what was moving (lips, eyebrows and so forth) but didn't keep resending parts of the image that didn't move (ears, say, or background). The compressed signal went through a few copper wires to Basalt, Colorado; by optical fibre to Denver; by a series of satellites to Perth, W.A.; by fibre again to Fremantle; by an improvised microwave link from the roof of the Telecom building to the roof of the conference hotel; by coaxial cable into the ballroom; through similar chips that spliced a high-quality, full-motion, full-colour picture seamlessly back together; into a video projector; and less than a quarter-second after a word was uttered in the mountains of western Colorado, there it was, perfectly synchronised with digitally echo-cancelled sound, on the retinas and eardrums of the audience in Fremantle.

Auctions for Second-hand Cars

North American car dealers typically travel as often as once a month to auctions where they examine and buy the cars they will resell. For dealers affiliated with Mitsubishi Motor Sales America, a 3- to 4-hour auction is held in six locations across the US. But since dealers are far more dispersed than that, with many in relatively remote areas, those not near one of the six sites may face up to 3 days' return travel for each auction. Now an interactive television system that has been in operation for 10 years promises to eliminate this heavy travel burden.

An online auction system called Aucu-Net, based in Atlanta, Georgia, has signed up 70–80 per cent of the 115 dealers approached to date for a system that makes auctions travel-free, cuts distribution costs, reduces product cycle time and improves inventory control. (Another 415 dealers, in three regions, are yet to be approached.) Participating dealers allow proprietary equipment to be installed at the dealership: a computer and digital receiver, colour screen, dedicated fax and satellite dish. Every Tuesday for 90 minutes, the dealer can use this dedicated equipment to view cars and their specifications and to bid with other online dealers. The dealers stay where they are; so do the cars until one is bought and shipped to them.

Dealers can also buy and sell cars in an open consignment auction held every Friday. Initial reluctance – some buyers at first thought they couldn't buy a car without being on the spot to see and touch it – has turned to eagerness to capture the vast savings in fuel, time and money. The programme is now being extended beyond the existing three regions. Meanwhile, all 530 dealers are saving paper through an electronic mail system from headquarters, where a paperless office system has been evolving since 1985.

A Factor of 100, but Perhaps Only 4

Returning to the standard question of this book, we may ask what amount of resources can be saved by telecommunication. Methodological questions arise over what to count and to compare. We opted for Schmidt-Bleek's MIPS method, the calculation of material inputs per service unit (see Chapter 9 and the 'introduction' to Chapter 2). Hartmut Stiller of the Wuppertal Institute and Thomas Egner of the Ulm-based FAW (Forschungsinstitut für anwendungsorientierte Wissensverarbeitung) obtained the following provisional results.

For a trans-Atlantic business trip, what has to be counted is the ecological rucksacks of the proportional air fuel consumption, of the aeroplanes, the stay in a business hotel and a few other items relating to the trip. As an estimate, the total ecological rucksack is about 1 tonne. A 6-hour videoconference, by contrast, may account for material inputs of rather less than 10 kg. This means that a half-day videoconference could represent (with a wide margin of arbitrariness) a MIPS reduction factor of roughly a hundredfold.

Clearly, these results have to be used with a great deal of caution. Not all business trips can be adequately replaced by videoconferences. Larger conferences derive much of their value from coffee-break chats, poster sessions, ad hoc discussions and delightful dinners, from side-programmes and from the initiation and renewal of personal friendships. None of this is susceptible to electronic data transmission. Moreover, e-mail and videoconferencing create their own growing demand and can even incite participants to plan additional journeys they would not otherwise have thought of. So whatever the mathematical results of the comparison between physical transportation and virtual or electronic transportation, the real-world reduction of resource consumption may be quite modest, rather in the vicinity of a factor of four.

On the other hand, the huge reduction potential may legitimately play a role if policies are adopted to 'make transport prices tell the ecological truth'; *i.e.* making transportation significantly more expensive (see, for example, Chapters 6 and 7). In that case, many people will more readily forgo certain trips, and take consolation in realising that much of the transport demand is, in reality, based not on need but on thoughtlessness.

The applications of videoconferencing are many and exciting. The marketing of special goods, even works of art, can be done via telecommunications. A popular chain of photocopying shops in the US is rapidly installing equipment so that someone in any city can hold a videoconference with someone in another city without prior arrangements. Tens of thousands of private installations also exist. For example, we recently discussed some technical apparatus by video with a colleague by courtesy of an art gallery that normally uses the equipment to show objets d'art to prospective purchasers. Even if there is no adequate equipment at the other end, you can always use your own equipment to produce a high-quality videotaped speech, send the videotape to the remote audience for playing on a standard video player and television, and handle the discussion by conference telephone. And simpler, lower-quality, but for many purposes quite acceptable, videoconferencing cards that fit into ordinary personal computers and use ordinary voice telephone lines are becoming increasingly affordable and easy to use.

Perhaps one of the biggest growth markets for telecommunications is 'telecommuting', which normally involves data exchange and could involve videoconferencing. Many office tasks can be done – and are increasingly being done – at a distance. Especially for families with small children, this represents a welcome opportunity to continue working without necessarily leaving home. To take another example, maintenance and repair do not always require the personal appearance of service experts. Much can be done via videoconferences. When RMI needed a certain electronic circuit board replaced from a Japanese manufacturer, for example, barriers of distance and language were instantly surmounted simply by placing the failed board under the graphics camera and pointing to the failed component so the supplier could see exactly what needed mending.

3.2 ELECTRONIC MAIL

The early exchange of manuscripts in writing this book was done by e-mail. Bits and pieces of the book were 'mailed' across the Atlantic from Snowmass, Colorado, to Wuppertal and Bonn, Germany. In addition, inquiries were made to Singapore, Brazil, Japan and other places using faxes, e-mail and telephone. Imagine how much time would have been lost and what amount of resources would have been consumed if the same exchange of messages and draft texts had been done by ordinary airmail. Maybe the production of the book's first draft would have taken two years instead of two months. It would have looked much 'maturer', no doubt, but it would have been outdated in many regards on the day of its publication.

How about the resource savings of substituting, say, facsimile for the physical transmission of mail? In Japan, where energy expert Haruki Tsuchiya has examined this question (1994), the entire telephone system uses energy equivalent to about 553 W for the duration of each telephone call. Japan's 56 million telephones (in 1991) made 74 thousand million calls totalling 3.4 thousand million hours or an average of 1,316 calls per telephone-year or 2.8 minutes per call. The energy consumption per call is thus 0.026 kWh, or the electricity generated by burning 12 g (0.4 oz) of coal. Clearly, if a phone call displaces a physical journey to talk face to face, it can save enormous amounts of energy. With a posted letter, however, the comparison is less dramatic: in Japan, where domestic post is moved by surface transport rather than by air, it takes twice as much energy to move a one-page letter by post as by a facsimile machine that is used only 5 times a day (since its 15-watt standby energy must be spread over those few messages), whereas if the fax machine is used, say, 50 times a day, it becomes about 92 per cent less energy-intensive than the post.

In a somewhat similar approach, Hartmut Stiller and Thomas Egner (see p 114) have estimated the average inputs for mailing a 10-g letter from Wuppertal, Germany, to Snowmass, Colorado. The MIPS 'weight' would be roughly a pound. This quantity represents the sum of crudely estimated proportional inputs to the paper manufacturing and to the surface and air transport services. The e-mail, on the other hand, causes no direct material inputs but is based on earlier material inputs that were required for building home computers, cables and satellites. Dividing those by the estimated average life spans of the hardware parts yields rough estimates for the MIPS of, for example, a 10-kilobyte letter. The result is some 5 g. The MIPS reduction factor would therefore be roughly 100. Depending on initial assumptions and the type of mailings one is comparing, the factor could just as well be 1,000 or only 20.

3.3 Strawberry Yoghurt

Germans are very fond of strawberry yoghurt. Some 3 billion cups of this delight are eaten annually in the country. Until the winter of 1992–93, no one had ever thought of strawberry yoghurt as more than an innocent treat. All this changed suddenly when findings from a study by Stefanie Böge made headlines. What Böge had established in early 1993 was the embarrassing, even ludicrous, transport intensity of the product. The yoghurt and its ingredients and the materials used for the glass cup made accumulated journeys totalling 3,500 km. Another 4,500 km could be added for the supplier's supply transports. Figure 11 shows Böge's findings on a map.

This map was carried by newspapers throughout Germany. Overnight, Stefanie Böge became the best-known researcher of the Wuppertal Institute, where she works in the transport division with Rudolf Petersen.

In a way, this sudden fame was a surprise to anyone with a basic understanding of modern manufacturing. Surely everyone knows that ores and metals and plastic and fruit are shipped around the globe, not only within central Europe. But with the beloved strawberry yoghurt, the German public cultivated the convenient illusion (in a country with abundant strawberry and milk production) that

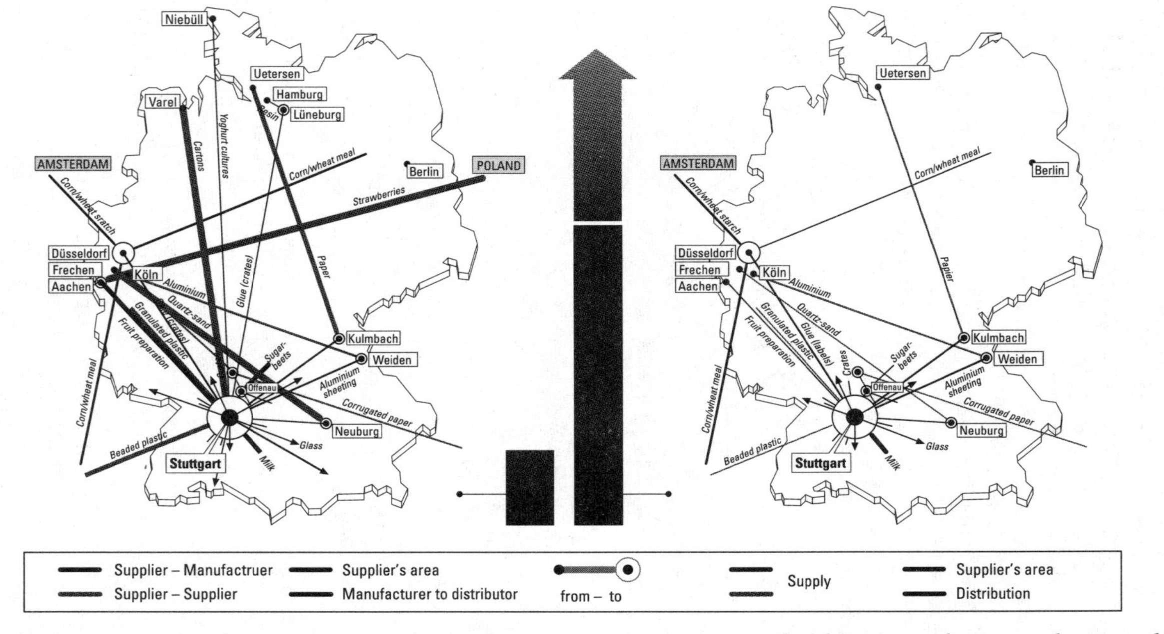

Figure 11 Thirty-five hundred kilometres is the average distance travelled by a strawberry yoghurt made in Stuttgart, Germany. It is not, of course, the finished cup of yoghurt itself that travels this extraordinary distance; rather, most of the distance is accounted for by the materials and ingredients that have to be shipped to the dairy manufacturer in the first place. The journeys of the supplies to the suppliers add another 4,500 kilometres. For comparison, the right-hand side of the illustration shows the geography of a low-transport-intensity yoghurt production. (After Böge, 1993)

the thing came from just round the corner. This is probably why Böge's findings met with such inordinate public interest.

Let us turn to the Factor Four question from here. Is it possible under today's commercial conditions to manufacture that cup of delicious strawberry yoghurt with a lot fewer kilometres?

Indeed it is. Strawberries, milk, sugar and other ingredients can be produced and processed locally. Glass jars can be returned in local or regional markets. For the lightweight aluminium lids, however, it would not make much sense to reduce the distance, as major investments would be required and not many lorry rides would be saved.

More savings could come from decentralised dairies. As consumers are becoming interested in the transport issue and are developing preferences for locally produced goods, markets for local food are being created or maintained (see the mushroom story, also from Stefanie Böge, at the end of this subchapter). One technical implication may be the installation at each production facility of the relevant equipment for hygiene, automatic filling, storage of processed fruit, packaging and so on. All this would have its price, economically and ecologically.

Much less investment would be involved if people were to revive home production. This is technically feasible, and the yoghurt is likely to be very tasty. Home production could, for all its technological inefficiency, be economically attractive, particularly for those families who emphasise the values of leisure time over participation in the money economy (see Chapter 14, p 294). They would tend to see such home production as a pastime rather than as an activity that must be technically optimised.

We seem to have learned that in this case Factor Four is not at all a technical problem. As a matter of fact, the suggestions made for reducing transport intensity imply rather less than more sophisticated machinery and technology. More human labour would be needed. In the case of home production, more 'dedication' might also be needed.

The Factor Four transition is not, however, going to happen under the market conditions prevailing in OECD countries – conditions characterised by expensive labour and cheap transport. It just does not pay to economise on tonnage. Yet for a country with high unemployment levels and with the need to import crude oil for petrol or diesel, reducing transport intensity while increasing labour intensity could prove profitable.

Marketing Low-Transport Produce: The Mushroom Story

When Stefanie Böge became widely known in Germany, food producers got nervous and asked themselves when their own products might become subject to such unwanted publicity. The German mushroom growers' association felt they should confront the issue head-on. They asked Böge to do a study for them in the context of their efforts in eco-management and -auditing.

On average, a 10-tonne truck drives 65 m in the service of producing a pound of white mushrooms. Horse manure is the single biggest contributor to this transport balance. Roughly 10 tonnes of it are required for producing a tonne of mushrooms. And German growers produce 58,000 tonnes of mushrooms annually. (We were surprised there were enough horses in Germany to supply what was needed for this quantity!)

In the growing beds, there is a thin cover layer of dark soil on top of the manure. That topsoil travels long distances and consists mostly of depletable peat. The mushroom growers have commissioned the experimental development of a cover consisting mostly of wastepaper. Jan Lelley from Krefeld succeeded in producing 'Champyros' (80 per cent wastepaper), which does the job marvellously and does not require long-distance hauling.

After scrutinising their own transport intensity, the German mushroom producers went on to publicise the relatively small distance their produce is travelling. This seems to be the first recorded instance of a business's actively advertising short distances in order to attract environmentally conscious customers.

3.4 Local Black-Currant Juice or Overseas Orange Juice?

Germans have become world champions in drinking orange juice – not because the German climate is any good for growing oranges, but because they just like orange juice. It's cheap and it's healthy, so why shouldn't they drink some 1.5 thousand million litres of the stuff (some 20 litres per person) annually? We are not saying they shouldn't. But we want to show that less transport-intensive alternatives exist.

Growing the oranges to produce those 1.5 thousand million litres of juice requires an area as large as Saarland, one of the 16

German states. Orange juice consumption therefore contributes considerably to the size of German 'ecological footprints' (see William Rees, Chapter 8, p 220). In addition, to transport the concentrates of the orange juice, some 40 million litres of fuel oil are consumed, producing more than a hundred thousand tonnes of CO_2 emissions.

The 1.5 thousand million litres of orange juice that Germans now drink every year represent an increase of more than 100-fold over the consumption levels of 1950. Local beverages including black-currant juice (with higher vitamin content than orange juice) have lost market share over the years. In 1965 three times more black-currant juice was marketed than today, in addition to very large quantities from home gardens not entering the market at all. By switching back from orange juice to black-currant juice, a factor of ten could easily be gained in 'transport efficiency', while also obtaining a considerable increase, perhaps a factor of two, in 'area productivity', *i.e.* the amount of beverage per acre. For further technical information see Kranendonk and Bringezu (1993).

Examples like this one do not connote a technological efficiency revolution. They rather convey the message to Northern readers that a degree of self-sufficiency may be needed to rediscover the pleasures of eating and drinking home-grown delights.

3.5 Quadrupling the Capacity of Existing Railways

Horror scenarios abound about the increase of traffic on European highways. The Single Market, enlarged in 1995 to 15 member states, is projected to lead to a doubling of cross-border goods transport from 1990 to 2010. The fall of the Iron Curtain has added an East–West dimension to the perennial congestion. For truck drivers, East–West traffic has become a nightmare. They regularly spend many hours, sometimes more than a day, waiting at customs between Poland and Germany. Road construction is costly and slow and meets understandable resistance, particularly in the crowded West.

Can rail be the answer? New tracks take 15 years to be planned and built, at horrendous cost, including environmental cost. And their tonnage capacity is normally only 50 per cent of that of a four-lane motorway.

But could we not revolutionise the tonnage capacity of the rail-

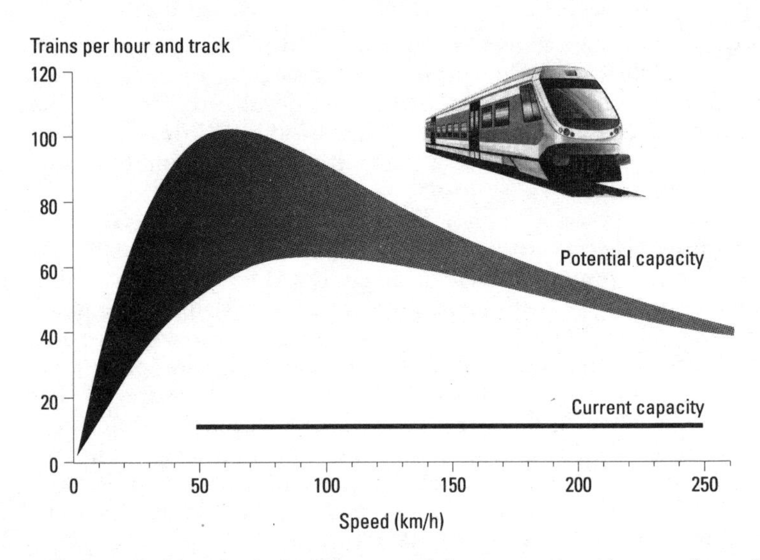

Figure 12 The theoretical potential for increasing the number of trains that can safely pass over one railway track. The straight horizontal line shows the current capacity. The graph illustrates the quadrupling of the capacity of the existing railway system that can be achieved.

ways? Yes, says Professor Rolf Kracke of the University of Hanover. He masterminded the 1990 'intelligent rail' concept (Kracke, 1990) and is presently elaborating his ideas in a major new study for the privatised German railway company Bahn AG ('Rail plc').

The main element of Kracke's proposal is safely increasing both the frequency of trains on the lines and the load capacity of freight trains. Today, the distance between moving trains is roughly 3 km, depending on velocity. This is because, under worst-case assumptions and depending on velocity and the signal system, it takes 3–5 km for a train to come to a complete stop after the first signal to stop. Kracke and his team are making proposals for new electronic control techniques for safely minimising the safety distance. Figure 12 shows the theoretical potential of increasing the transport capacity of existing railway tracks for the full range of velocities.

The capacity is not only dependent on the tracks laid through the open countryside. Terminals must be improved too, but not expanded in size. As a matter of fact, the 19th-century-technology shunting stations are far too large. They are woefully retrograde in terms of both land use and technological advancement. Modern switching can be done by horizontally moving just containers across a platform, not entire railway cars. Twenty or more containers can simultaneously be switched from one train to the other or into a big warehouse. Using such simply sophisticated means, an entire goods train can be reconstituted in a mere 15 minutes.

3.6 Pendolino and CyberTran: The Soft Options for Rapid Trains

As stated in the preceding subchapter, the railway is generally preferable to road and air as regards resource use per passenger- or tonne-kilometre. Moreover, rail passengers can enjoy a comfortable working space, a nap or a restaurant lunch while travelling – conveniences not available when driving their cars. The disadvantage of trains as opposed to cars is, of course, a lack of flexibility in the local area. The lower limit of distance where trains are competitive with individual cars may be in the range of 50–100 km. The upper limit of competitiveness where trains have to compete with aeroplanes may be in the range of 400 kilometres. That distance can be expanded for higher-speed trains. So modern high-speed trains – French train à grande vitesse, TGV, the Japanese Shinkansen and the German ICE (Inter-City Express) – have become the favourite mode of transport for business and private journeys at distances of up to some 800 km. Germany is planning to build Transrapid, a magnetically levitated train that is designed to 'fly' at 500 km/h.

The trouble is that Transrapid is very dear and will, of course, create horrendous noise at that high speed. Environmentalists are not at all convinced that Transrapid is environmentally any more desirable than road or air transport. Their assessment of TGV or ICE is not much more favourable, because those conventional high-speed trains require straight tracks that brutally cut through the landscape and that, again, are very costly.

Luckily, a better solution is available for fast trains. It is called Pendolino and was invented by Italian engineers. On curvy tracks, it will tilt in a way that allows it to maintain its high speed. The typical travelling speed will be some 150 km/h, even on most of the curvy tracks that now exist in European countries. The investment, therefore, will be much smaller than in the case of TGV/ICE, let alone Transrapid, and yet the range of competitiveness for the rail will be nearly doubled. Fortunately, several European railways, including the privatised German Bahn AG, are heavily investing in the Pendolino technology, which seems easily to meet the Factor Four criterion when compared with passenger cars or air transport.

CYBERTRAN

If an entirely new system should be designed, it ought to be much more resource-efficient than the existing ones. One such innovation has come from the US. The Advanced Transportation Systems group at the Idaho National Engineering Laboratory (INEL, Idaho Falls, Idaho) has developed a working prototype for a transportation system that uses one-tenth the fuel per person of automobiles or airliners. It also costs five- to perhaps tenfold less to build per kilometre than a highway or a railroad, and would cost the traveller significantly less to use than buses, planes, trains and cars. The name of this remarkable innovation is CyberTran (Dearian and Plum, 1993). Plate 10 shows the model.

CyberTran is a computer-controlled (*i.e.* no-driver), low-occupancy, ultralight rail vehicle. Each car weighs 10,000 lb (one-tenth the weight of conventional trains), including a full complement of 14 passengers. (Denser seating can accommodate up to 32 passengers.) CyberTran is propelled by two 75-kW electric motors at speeds of up to 150 mph on an elevated guideway. Its steel wheels rest on two simple steel pipes, each welded to a horizontal steel plate; the friction on the small contact patch is high enough that CyberTran can climb steep grades. The elevated guideway is so narrow that it can be retrofitted above the median of an existing road. If installed from scratch, no road need be built beneath it to provide access for its construction.

Perhaps most interestingly, CyberTran is an on-demand transit system, which means that it runs only when a traveller desires it (like cars and elevators), and it runs as directly as possible to the traveller's destination. In contrast, all conventional mass transit systems run only at scheduled times or periodically. Whether the CyberTran passenger load is low or high, there are only as many vehicles in use as are needed. During periods of low use, parked vehicles are distributed throughout the rail system, ready for immediate service to potential passengers (for example, every half-mile). So compared with periodic or scheduled-time transit systems (such as buses and aeroplanes) that run at times of low occupancy, CyberTran is more than tenfold more energy-efficient without sacrificing service to travellers. It is also more economical than an on-demand bus system would be, because individual human operators are not needed in the CyberTran system.

The modest size and weight of an individual CyberTran vehicle account for its significantly lower capital costs and per capita energy consumption. In a new rail system, usually between 70 per cent and 80 per cent of the capital cost is for building the roadbed, laying tracks, building bridges and installing power lines. These infrastructure costs are so high because the structures must be designed to support train cars that weigh 100,000 lb. Reducing vehicle weight by 90 per cent cuts the infrastructure cost for building a CyberTran system by more than tenfold. For example, the projected $2 million cost per mile to build a CyberTran system is roughly 87 per cent less than the cost of a new high-speed rail system. Also, the capital required per mile for CyberTran is a factor of five lower than the $10 million average cost per mile of building automotive highways.

CyberTran's lightweight design also makes it more energy-efficient compared with other modes of transport. It requires 10 per cent of the fuel per passenger-mile of a single-passenger automobile and 7 per cent of the fuel for a 60-per-cent-full Boeing 737 (the average passenger load for the US airline industry). CyberTran also uses far less fuel than standard high-speed rail, such as France's TGV. In fact, the TGV uses more energy per passenger than CyberTran unless the TGV is carrying its full 500-passenger load.

CyberTran's high energy efficiency and low cost do not come at the expense of convenience or service. For distances between 100 and 300 miles, CyberTran is just as quick as aeroplanes when you take into account the time it takes to get to the airport, board the plane and so on, and three to five times quicker than cars over the same distance. And unlike those modes, CyberTran can operate in bad weather.

The developers of CyberTran believe that the most useful near-term niche for their innovation is for intercity travel of distances between 100 and 500 miles. As mentioned above, applying CyberTran to this range is attractive, because it would be as quick as aeroplanes, and quicker than cars, at a much lower environmental and economic cost than either. However, INEL expects that CyberTran will also be more economical and efficient than most urban rail systems. Two exceptions are large, highly utilised metropolitan underground systems and high-speed rail systems that operate with consistently high load factors. Examples include most of the London Underground and the French TGV, respectively.

Who will be the first to take advantage of this most efficient and economical innovation remains to be seen. The city of Boise, Idaho, is considering a test installation, and several other municipalities are believed to be interested in the technology.

3.7 CURITIBA'S SURFACE UNDERGROUND

Curitiba is the capital of the state of Paraná in southern Brazil, 200 km from São Paulo. Its population of 1.6 million inhabitants has more than tripled in the past quarter-century, making Curitiba the fastest-growing city in Brazil. Yet despite its phenomenal growth rate, it is also one of the most liveable cities in Brazil, and possibly on the continent.

Underlying Curitiba's success is the city's 'master plan'. Developed in 1964, it defined an integrated transportation and land-use plan that structured the city on two axes, along which both commercial and residential development were promoted. Each axis is formed by three parallel roads. The central road is dedicated to mass transit, while the two outer roads are one-way for all other traffic. Before the plan was approved, the city bought large amounts of land along these axes and built low-income housing there, ensuring access for everyone to the centre of the city. Since 1964, three more axes of development have been added to the original two.

Integrated planning has been the clarion call of the city's development since soon after the master plan was set. Curitibans have had much success with innovative approaches to zoning, trash collection, welfare, education, transportation and even flood management.

One of the key components of that success is its transit system, pioneered by former mayor (now provincial governor) Jaime Lerner. In 1971, when Lerner became mayor of the city and assembled an exceptional group, chiefly of fellow architects, to rethink its needs and opportunities from first principles, the automobile was quickly becoming the dominant transport mode despite the master plan. Lerner recognised that this trend was limiting the city's accessibility and took decisive action to counter the automobile's dominance. Building an Underground system was not affordable, so Lerner developed an alternative system with the hope that it would provide similar performance – but at a 500-fold lower capital cost.

Lerner implemented a network of buses that use the dedicated mass transit lanes along the axes and that feed people from other areas to the axes. Since its inception, use has skyrocketed, and the system has undergone many radical changes to accommodate its increased use from 50,000 passengers per day in 1974 to more than 800,000 passengers per day in 1994. These changes have improved the system until it can now handle more than four times the throughput of conventional bus systems (for more details see Rabinovich and Leitman, 1996).

The first factor that improved throughput was exclusive bus lanes. These roughly doubled the system's capacity. As demand increased and transit planners looked for further improvements, they chose to introduce longer buses that are articulated – hinged in the centre so they can bend round turns. This increased throughput to two and a half times that of conventional bus systems.

The next refinement was unique. To make the buses move people faster, Lerner's team invented tube stations – kerbside tubular glass-and-steel bus stations, closed at one end, with raised loading platforms. Passengers pay their fares as they enter the tube's open end. This speeds boarding, because the passenger needn't take the time to pay the fare on board the bus itself, and all of the bus doors can be opened for boarding. Thus, when the bus pulls up alongside the tube, big sets of matching doors open in the facing sides of both the bus and the tube, passengers board just as quickly as they would board an Underground train and the bus pulls away – with buses arriving as frequently as one per minute during the rush hour. Entry to the bus is also faster (and wheelchair-accessible) due to the level boarding platform, and the bus needs no ticket taker. Tube stations improved throughput to 3.2 times that of conventional busing.

The latest technological improvement has been the introduction of a biarticulated bus, yielding a transit system whose throughput is four times higher than conventional busing (see Plate 11).

The Curitiban transit authorities have also refined the route system by adding express buses and by building 20 transfer stations which connect axis routes with circumferential routes and with routes from the city's outskirts. They have also adopted a flat 'social fare' equivalent to 20 pence, valid for unlimited transfers. This fare structure was chosen so as not to penalise those living in the poorer outskirts of the city.

For anyone who has travelled on the Underground in any major city in a 'developed' country, this fare sounds very low, and it is. But remarkably, Curitiba's bus system is unsubsidised. Fares completely cover the costs of running the system, which is jointly run by private companies and the city. The city builds and maintains the infrastructure such as roads, transfer stations and tube stations, while private companies own and operate the buses and the fare collection system under license from the city. The private companies are paid per kilometre of bus route that they operate, not per passenger. This has encouraged extensive coverage and over 500 km of bus routes throughout the city and its environs.

Looking beyond its uniqueness and inherent effectiveness, the real importance of Curitiba's bus system is how well it is embraced by the city's inhabitants and how many knock-on benefits it produces. Nearly 70 per cent of the population uses the bus system each day. Benefits include a per capita petrol consumption 30 per cent lower than cities of comparable size in Brazil, the cleanest ambient air in any Brazilian city, the highest car ownership and the lowest car drivership. As a part of the integrated urban development plan, the bus system has also helped the city to achieve 52 sq m of open space per capita – higher than any city worldwide. Combined with other innovative educational initiatives (which use old buses as mobile classrooms, clinics and vocational libraries) and trash collection (integrated with recycling and nutritional programmes), the city is demonstrating an ever-growing understanding of system-based problem solving from which all other cities could learn.

3.8 CAR SHARING IN BERLIN

Some people in our cities simply haven't the means to own a private car. Some would have no place to park it anyway. Some others have ecological motives for not wanting to own a car. And some think very practically and consider it a nuisance to own a car if there is another choice available for daily errands and commuting. Now a choice exists in an increasing number of European cities: car sharing. Several hundred individuals in a city or a smaller neighbourhood get together and buy a few dozen cars which belong to all and which are accessible to all.

For joining the carpool, every member has to pay an entrance fee of DM 1000 (about £400). Additionally, each member invests an ownership share of another DM 1000 and pays an annual membership fee of DM 120 (about £50). Also, a distance and fuel fee of 52 Pfennig (about 20 pence) per kilometre and a user time fee of DM 3.90 (about £1.60) per hour are due. These fees pay for the fleet on a nonprofit basis.

Markus Petersen (1994) investigated the effects of car sharing in Berlin. In Berlin they call it Stattauto. Statt means 'instead' but is also a pun on the German word for city, Stadt. Stattauto has become very popular in this largest and most populous German city. First, Petersen wanted to know how many members owned a car until they joined the car-sharing pool. Only 21 per cent did. The others had never owned a car before (7 per cent), or were sharing a car on

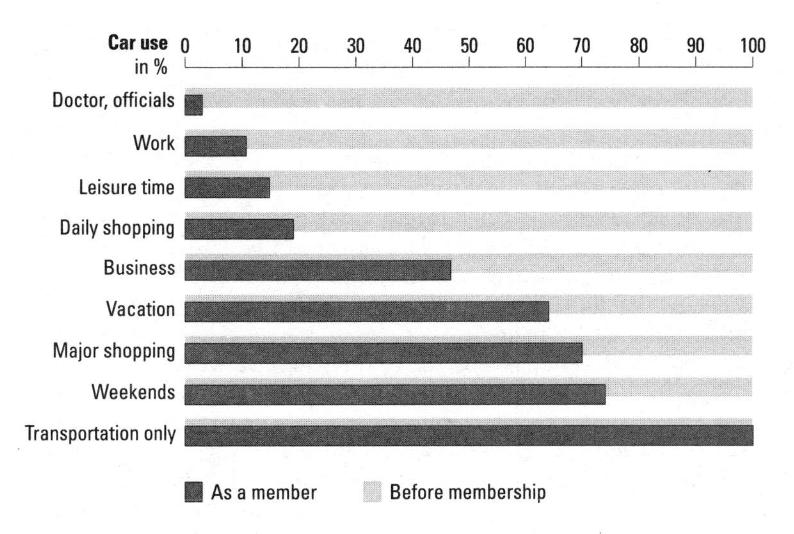

Figure 13 Car sharers make less use of a motor car than private owners of cars. The reduction of car use, however, depends on the functions performed. The graph shows how car use changed when people joined the Stattauto group, for nine different functions. Note that commuting to work by car came down by nearly 90 per cent. (From Petersen, 1994, p 192)

a private basis with friends or relatives (25 per cent), or had not owned a car for some time prior to joining the pool (43 per cent). After joining, half of the group members reported having no cars available other than the vehicles in the pool.

Car sharers showed a high degree of satisfaction with the service rendered. For most of them, mobility increased. Monthly costs were cut for those previously owning a car. Many also felt a relief from not fearing damage to or theft of their cars. Commuting to work by car was drastically reduced when people joined the group. Bicycle and public transport use rose correspondingly. Figure 13 shows the reduction of car use according to functions.

From the Factor Four perspective it is important to learn how much was saved by the scheme. According to Petersen, the creation of Stattauto reduced car ownership by 105 cars (51 people sold their cars when entering, 54 decided to join instead of buying a car). On the other hand, 27 Stattautos were bought. This means that the actual number of cars bought shrank from 105 to 27, which comes close to a factor of four.

Also, the total auto mileage was reduced, but only by a factor of two, which means that Stattautos were driven more than privately owned cars – more service per unit investment, materials, and space.

It can be assumed that the reduction of auto miles has to do with the price signal given for each additional mile. Ordinary car owners 'see' the costs for owning a car only when buying a new one. To them the additional mile is only its fuel cost. Fuel tends to be only one-fifth or less (in the US, nearer one-eighth) of the average total cost per mile, which includes depreciation, insurance, taxes, maintenance and repair. All those standing costs have to be borne on an equal sharing basis by the car-sharing participants.

Car owners would receive a more realistic feedback if they, too, were paying the average total cost per mile on a per-extra-mile basis. Theoretically, this could be engineered. If auto manufacturers were leasing rather than selling their products, and if the better part of the leasing fee were paid on a per-extra-mile basis, those per-mile fees would lie in the vicinity of one DM (about 44 pence). As a consequence, the car user would have a noticeable incentive for using his or her car only when needed. Imagine such an incentive structure's becoming a mass phenomenon: in a city like Berlin an additional million potential customers would be created for mass transit systems, both public and private. Mass transit would receive a powerful push forward and would be in a position to expand, renovate and even make profits.

3.9 Car-Free Mobility

Most people would warmly welcome superefficient cars for personal and environmental reasons. Yet there are more exciting perspectives for those who want to live in a more ecologically sustainable way. Why not try and live without an auto altogether? Exactly this has been a plan for some 200 families in Bremen, a North German city of half a million inhabitants.

The city of Bremen has taken a courageous initiative for a car-free city district (Krämer-Badoni, 1994). Construction began in autumn 1995 in Bremen-Hollerland of a new complex of apartments and homes which may be rented or acquired only by families renouncing the possession of a car.

The idea of car-free mobility originated not so much from ecological as from quality-of-life considerations. Car-free life under favourable conditions is often experienced as being very

Plate 1 A tropical garden at 7,100 ft in the Rocky Mountains. Outside there is snow. Yet inside, two dozen banana crops have flourished with no heating system. The house, which serves as Rocky Mountain Institute's headquarters, sells surplus homemade electricity back to the utility.

Plate 2 The Darmstadt Passivhaus needs 90% less heat and 75% less electricity than normal German houses the same size, yet comfort is superior.

Plate 3 De Montfort University in Leicester is proud of its new engineering building, opened by Her Majesty the Queen and designed by Alan Short and Brian Ford. The naturally cooled and ventilated building uses only 25–50% of the energy typically used in buildings of this type, but costs less.

Plate 4 ING Bank's headquarters in Amsterdam is 12 times more energy-efficient than its predecessor building, and much more agreeable too. The building, designed by Anton Alberts, has a lot to do with ING's fairy-tale-like success story.

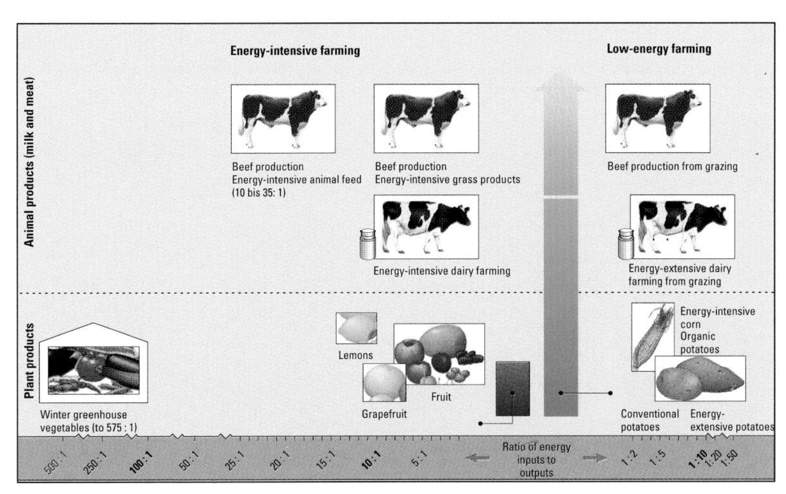

Plate 5 Farm produce has become a net energy consumer. Traditional farming (on the right side of the illustration) had a favourable input/output relation for energy. Beef from European mass producers can require as much as 35 calories of energy input per calorie contained in the final product (upper left corner). (After Lünzer, 1992)

Plate 6 Lee Eng Lock of Singapore is an efficiency wizard at air-conditioning systems, water pumps and cleanrooms. His motto: 'Like Chinese cooking. Use everything. Eat the feet.' (© Supersymmetry Services Pte Ltd)

Plate 7 One CD-ROM can replace uncounted files. A MIPS reduction by a factor of 50 may be attained. And the information is more conveniently accessible too.

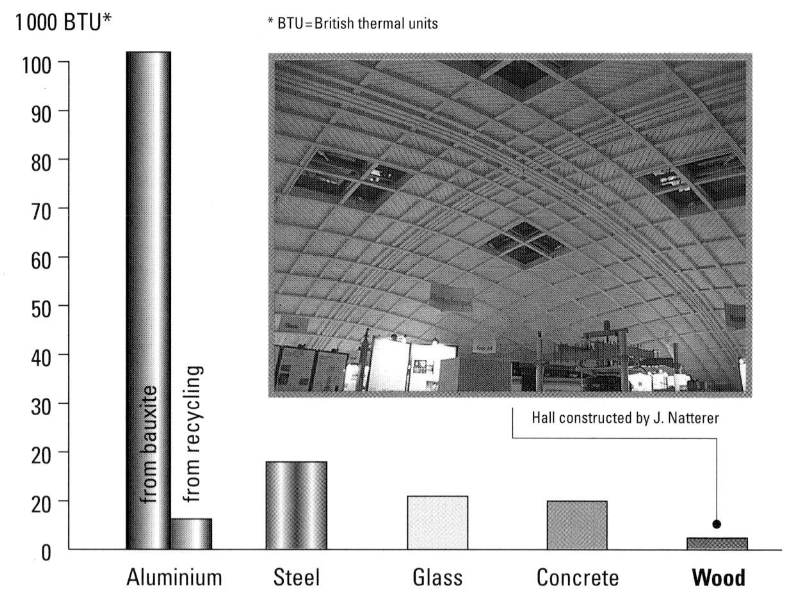

Plate 8 Wood is extremely versatile and useful as a construction material. In energy use, it compares favourably with aluminium, steel, glass and concrete (Browning and Barnett, 1995). The insert shows that contemporary widespan constructions are also possible with wood, as demonstrated here by the wood engineer Julius Natterer. The picture shows the trade fair hall of Ecublens on Lake Geneva, architect Dan Badic.

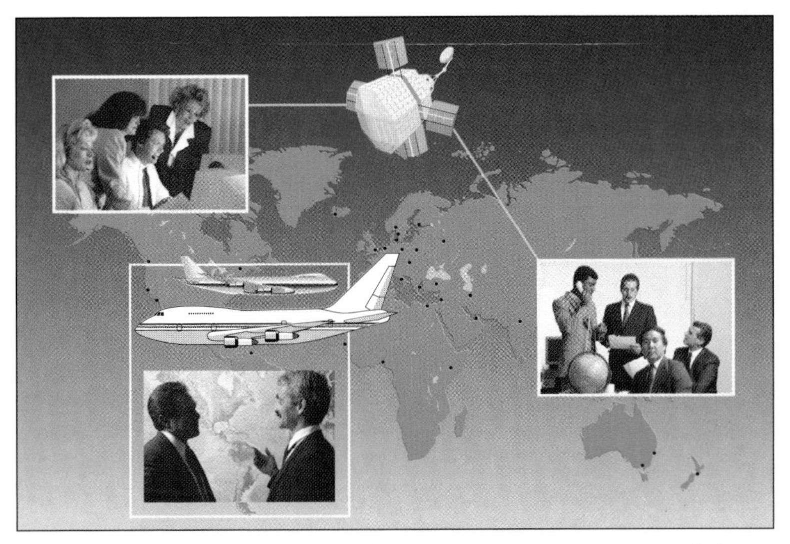

Plate 9 A six-hour videoconference can save some 99% of the energy and material resources that would be consumed by the transatlantic trips required to hold the same meeting in a single location.

Plate 10 The next generation of trains may come from the country which neglected trains most: the USA. It could be the Idaho-built CyberTran. It is a lightweight, high-speed train travelling on an elevated guideway than can be installed above existing roads at about 10–20% of the normal cost. And it's energy-efficient too.

Plate 11 Curitiba's unsubsidised buses are owned by private companies paid per mile of bus route, not per passenger. Nearly 70 per cent of all Curitibans regularly use the bus system! The kerb-side 'tube stations' enable one- to three-section buses to stop at one-minute intervals, achieving Underground-like passenger capacity at about 500-fold lower capital cost.

Plate 12 Feedback helps to make managers responsible. Imagine what happens to water purification at the plant if effluent from the plant is piped back into its water intake.

Plate 13 At Carnoules, in the South of France, the Factor Ten Club was founded in October 1994. The picture shows, in the front row from left to right, Heinrich Wohlmeyer, Franz Lehner, Friedrich Schmidt-Bleek (the Club's initiator), Richard Sandbrook, Miki Goto; second row from left: Walter Stahel, three aides, Ashok Khosla, Wolfgang Sachs; third row: Bob Ayres (behind Ashok Khosla) and Leo Jansen. Standing: Wouter van Dieren and Hugh Faulkner.

*Plate 14
A Galápagos Darwin finch using a cactus spine for picking insects. Had there been a land connection between South America and the islands, the insect-picking niche would no doubt have been defended and maintained by woodpeckers.
(© Oxford Scientific Films)*

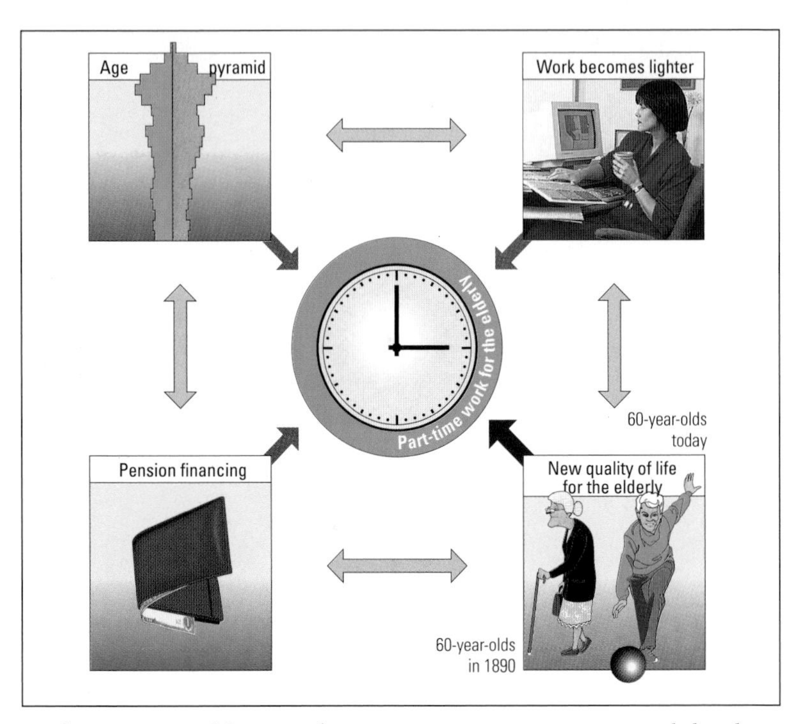

Plate 15 Nonobligatory but remunerative part-time work by the elderly can simultaneously solve problems posed by changing demographics (upper left corner), financing tomorrow's pensions (lower left) and the desire for an active new life cycle (lower right). The changing quality of work, lighter and more experience-based (upper right), makes re-entry into the job market of the elderly not only tolerable but highly desirable. (Design by Hans Kretschmer after the scheme of Reday-Mulvey presented in Giarini and Stahel, 1993, p 136)

agreeable, even when it is not based on any deliberate decision, but just on the lack of means to buy a car. The favourability of conditions tends to relate to the location of kindergartens, schools and shops, to public transport quality and to the recreational quality of the neighbourhood.

But car-free citizens have never had much opportunity to talk about their experiences and about the conditions for an attractive car-free life. City planners have traditionally seen the lack of a car as a grave deficit. Building and planning codes require space for cars, and streets wide enough for parking on one or both sides and for through traffic. So why bother about eccentric and long since outmoded car-free lifestyles?

The Social Democratic and Green Party aldermen then in power in Bremen felt they should listen to and acknowledge the experience of car-free citizens. They felt that the conditions should be made to favour citizens who were costing the community much less than average in terms of space and road construction and maintenance. Accordingly, the future dwellers of Hollerland were invited to participate in the local planning process.

In the end, however, the invitation to move into Hollerland was not taken up by a sufficient number of citizens. This lack of widespread acceptance was related to a number of factors: the higher than average price of the flats, the growing oversupply of housing in Bremen, the weak economy and the not so attractive location of the new development. The most significant barrier, however, was the uncompromising demand from the citizens involved in the planning process that renouncing car possession be binding not only on themselves but also on any potential future landlord or tenant. Making this stipulation legally binding in the land register significantly reduced the market value of the apartments and made the purchase a poor bargain. Even if the first owners were fully committed to life without a car, they did not want to lose a lot of money if and when they decided to sell their flats. It is now expected that the stipulation will be softened so as to attract a sufficient number of citizens to make the programme viable.

In order to make the car-free city a real success, compromises will be sought. Obviously, fire engines, ambulances, taxis and delivery vehicles will have access at any time. Rented cars or shared cars (see 'Car Sharing in Berlin', above) may be used for special missions and for holidays. And the restriction on future landlords will be softened. Nevertheless, living in Bremen-Hollerland will mean attaining much more than a factor of four in cutting transport intensity.

When word got around about the Bremen-Hollerland plans, more than 40 other German municipalities became interested and began to develop similar plans. Among them are the cities of Nürnberg and Freiburg. Part of the motive for this widespread interest was a desire to assist the less well off who just cannot afford a car. The transport intensity of the inhabitants of an 'autofree' city district will on average decline by more than a factor of four. The compensatory increase in travel by tram or bus will necessarily use fewer resources than the displaced car driving.

A somewhat more comprehensive – and more successful – approach, attaining much more than a Factor Four in resource productivity and drastically reducing transport needs, was taken in Bielefeld. This project goes far beyond city life with no private cars. Dwellers of the 130 new apartments at Bielefeld-Waldquelle will use some 70 per cent less water than average, compost all organic wastes including human wastes, live in houses built from local materials (mostly wood and clay-based bricks and roofing tiles) and obtain most of their food from organic farms in the neighbourhood. Hans-Friedrich Bültmann, the initiator, also arranged the planning of a trade centre inside the Waldquelle 'village'. Some 200–300 jobs in a great multitude of trades will allow most of the dwellers to commute to work on foot. Also, attractive recreational facilities are to be found in or very close to the village, including a newly provided little lake.

3.10 Getting the Village Feeling in the City

In his 1989 book *A Vision of Britain,* His Royal Highness, the Prince of Wales wrote: 'I am hoping we can encourage the development of "urban villages" in order to reintroduce human scale, intimacy and a vibrant street life ... that can help to restore to people their sense of belonging and pride in their own particular surroundings.' A small group of designers around the world is doing exactly that – and in the process is greatly reducing the need for motorised transport in our everyday lives.

After almost a half-century of designing neighbourhoods in which people are hardly neighbourly – zones designed for cars, not for people – new patterns of humane housing are re-emerging. Architects and developers are realising that clustered housing, narrower streets for reduced speed and noise, usable open space, and preserved or restored natural areas can add not only aesthetic but also major economic value. This approach is often labelled 'neo-traditional' or 'pedestrian pocket' design. The pedestrian pocket is thought to be the most energy-efficient form for a community, and conducive to creating a social community as well. Environmentally minded, market-conscious designers are increasingly creating houses, neighbourhoods and entire communities that integrate homes with workplaces, food production and the natural environment, all within easy reach of each other and often intermingled.

A major driving force for these developments is the infrastructure and land cost of roads and utilities. The best-known of many US comparisons of conventional densities with higher or mixed densities and clustered forms, a multiagency federal government study called *The Costs of Sprawl,* found that on a given land area, a high-density planned development could leave over half its land area as open space and significantly reduce road and utility investments, compared with a traditional suburban layout. Reduced paving would also reduce storm runoff, and shorter distances would reduce automotive fuel use and air pollution – as could, if desired, clustering and attaching some homes to reduce the area of exterior walls. The analysed comparison found a 35 per cent, or $4,600 (1987 US$), lower cost per house for site preparation, roads (20 instead of 30 ft wide), driveways, street trees, sewers, water service and drainage (done by natural swales instead of full kerbs and gutters).

These findings have been richly borne out in practice, even in autocentric America, by successful commercial developers such as Michael Corbett. His Village Homes project in environmentally minded Davis (near Sacramento), California, was built from the mid-1970s to the early 1980s, growing to 200 buildings on 70 acres. Mixed housing types on narrower streets, greenbelts with fruit trees, agricultural zones in among the houses, natural surface drainage, solar orientation and abundant open space created a delightful ambience. Like two interlocking sets of fingers, two separated access networks served each house: pedestrian access from the common greenbelt across the front, and a carport linked to a 20- or 24-ft-wide tree-shaded street across the rear. Walking and emergency-vehicle access are protected by a 3-ft easement on each side on which nothing can be built nor plantings allowed to exceed 6 inches high. With plentiful social interactions, the crime rate was only about one-tenth that of other developments nearby.

Most travel is by walking and bicycling in the car-free greenbelts. The bike paths connect to bike lanes on the streets. People are allowed to conduct business in their homes (illegal in many

American communities). Wider issues of colocating jobs and housing were little addressed in this early project, and there is only one small commercial building in the project. However, a strong emphasis on suburban farming and forestry (fruit and nut trees) makes this perhaps the only American subdivision noted for the quality of its organic vegetables, and able to finance much of its parkland maintenance from selling its almond crop.

The narrower streets not only calm traffic and save money and land but also save paving material and improve the summer microclimate, since trees can shade the whole street, and the expanse of dark paving to absorb and reradiate solar heat is reduced. Recent data confirm that narrower, tree-shaded streets in California's Central Valley communities could lower ambient summer temperatures by 10–15F° (6–8C°) throughout the whole area, greatly reducing air-conditioning energy.

Using natural drainage swales instead of costly underground concrete drains saved $800 of investment per house, nearly enough to pay for the landscaping of the parks and greenbelts. But the biggest economic benefit has been the project's eager market acceptance. Per unit of floor space, Village Homes, originally modest in their market positioning, now realise the highest resale prices in Davis, sell much faster (when they even get listed for sale – most are snapped up first by word of mouth), and fetch $15–20/sq ft (£110–140) more than homes in surrounding projects. Although it is difficult to separate the pedestrian-oriented features of the project from other 'green' features, they are certainly important to its exceptional financial performance.

Further confirmation of the commercial benefits of focusing suburban design on people and limiting the role of cars comes from the 400-ha Laguna West development in Sacramento, whose first model houses opened in 1991. Redesigned from conventional suburbia and strip-mall plans to a 'pedestrian pocket' style by architect Peter Calthorpe and developer Phil Angelides, Laguna West will integrate parks, lakes, commercial, retail and industrial space with its more than 3,000 homes. Its houses have front porches facing the street. The garage is moved discreetly to the back – an initially controversial break with the previous American norm of having two or three garage doors facing the street, as if (says architect Andres Duany) to announce, 'Cars live here'. Rather, the streets and common areas are made inviting for walking and biking, and transport emphasises public modes and car sharing. As at Village Homes, tree-shaded streets, narrowed by planting trees in the normal parking lanes, cut summer temperatures and invite people outdoors.

The narrower streets were at first unacceptable to the authorities: agencies operating fire trucks, ambulances and garbage trucks all had different width requirements that were added, yielding standard streets 'wide enough to land a plane on. So,' said River West Developments' Susan Beltake, 'we set up a demonstration street [demonstrating the proposed, greatly reduced width] in one of our neighbouring developments and videotaped. We brought in [potted] trees and concrete barriers for the in-street tree wells. All our employees' cars were parked on the street. We rented an ambulance, garbage truck and a fire truck that can handle up to a seven-storey [high-building] fire. (The tallest building in the area will be three storeys.) We drove [all these vehicles] ... up and down the street [and filmed it] .... It all cost about $5,000.' They got their permission, leading the way for other developers to cite their precedent.

Marketed as 'an Old Fashioned New Community,' Laguna West is recreating nearly forgotten behaviour patterns centred around front porches, streets and the central village green with its park, town hall, transit facilities, library and day-care centre. (None of this would seem unusual in traditional European towns, but in America it had been all but forgotten over the past 40 years.) Most of the 1,858 single-family homes are set around a 30-ha lake; and off of two radial boulevards is a lower-school site and more park space. Inspired by a modified beaux-arts system, this layout complements streetscapes dominated by human activities, not cars. Calthorpe notes that pedestrians 'want safe, interesting and comfortable streets to walk on: tree-shaded, with houses and shops fronting directly on them for interest and security. They want detail and human scale in the edges and places of a community. And they want narrow streets lined with entries and porches leading to local shops, schools, parks, not curving streets lined with garage doors leading to six-lane arterials.' Market response is reportedly strong. And development cost was identical, but for an extra $800 per house for the lake and $700 for the street trees – features that added an order of magnitude more real estate value than they cost.

A third example of pedestrian-oriented development, Haymount, is a new town being created by the John Clark Company on 1,700 acres along the Parrahanock River in Virginia. Planned by Andres Duany, the site will ultimately house 12,000 people in 4,400 clustered homes, with over 60 per cent of the site left undisturbed. Haymount is the first development on this scale to address energy efficiency, sustainable materials, habitat restoration and conservation (including detailed species mapping), and biological wastewater treatment. The exceptionally attractive site plan includes a very fine-grained neotraditional mixture of houses, retail, commercial, light industrial and agricultural sites, as well as parks and natural and restored areas, 14 sites for houses of worship and a transit centre. Businesses have been prerecruited to provide local jobs, lest residents commute 40 minutes by light rail to Washington, DC, defeating the intent to create a largely self-contained community. All homes are within a 5-minute walk of the village centre and a shuttle stop. A citizen-based planning process has built much excitement and deep community involvement in the design. Construction should begin shortly; the prospects for commercial success are already bright.

It is far too early to suppose that these three illustrative projects will rapidly transform deeply rooted American habits of ceding community design to road engineers. But their popular acceptance and favourable economics show that the opportunities they create for 'negacars' and 'negatrips' can be welcome both to developers' bottom lines and to the yearnings of those who increasingly see themselves, as architect William McDonough puts it, as 'people with lives, not consumers with lifestyles.'

Part II

Making it Happen: Improving Profitability

Introduction

Part I makes for exciting reading, we feel. It really is like discovering an entire new continent! Great business opportunities are waiting to be seized. The economic value involved in implementing those 50 examples of the expected efficiency revolution on a worldwide scale could be immense.

We cannot help comparing these opportunities with those relating to genetic engineering. Thousands of business analysts are scrutinising this high-tech field – only to discover one disappointment after another. On the other hand, no equity analysts seem to be looking at the new universe of efficiency technologies. Something seems to be wrong.

What is the problem? Are our examples technologically incorrect? Hardly. We feel, in fact, that they are much sounder, all of them, than most of the biotech promises that have attracted so much attention and venture capital.

Another explanation could be that many of the examples given seem to run counter to market expansion. Think, for example, of low-energy beef, durable furniture or the car-free city district. Shrinking turnover may repel rather than attract capital that is seeking profitable investment. There is some truth to this interpretation, as we shall be discussing in Chapter 12. Some of the efficiency gains are indeed achieved by doing away with nonsensical turnover. But that would not in itself mean that capital cannot be interested. After all, it was the business world itself which moved from the old-fashioned obsession with turnover to the lean and profitable firm, recognising that what matters is not the top line (total revenue) but the bottom line (net profit). More important, the better part of the examples are fully compatible with market expansion – to meet the needs of thousands of millions of people under the conditions of limited resources.

A different explanation could be a lack of purchasing power for efficiency technologies. Even if the need for efficiency is well established, as we firmly believe is the case, there just may not be enough purchasing power behind that need to create a sufficient demand in the marketplace. The analogy that comes to mind is the well-known problem of finding a cure for a tropical disease typically afflicting the poor; nobody would dare to deny the need for the cure, but it still might not be profitable enough to justify major investments on the part of pharmaceutical firms (in which case the World Health Organisation or some charity trust might be willing to help out).

However, the 'tropical disease' explanation is not satisfactory either. We are not talking about a problem limited to a certain population or geographic area or about something charities are concerned with. It is the world *problematique* itself, in the language of the Club of Rome, that we are covering in this book. We shall be demonstrating in Part III that the Factor Four revolution will be at the heart of the solution to the secular challenges that were voiced in *The Limits to Growth* and that led to the Earth Summit of Rio de Janeiro.

The new technological revolution is urgently needed and is technologically realistic. What else could be missing? It comes down to two drivers of capital investment in the real world: information and, in many cases, profitability. Of course, most of the examples given are profitable – in the sense that the investments will pay off after a certain amount of time. But they have to compete for attracting scarce capital with other investments that are simply – or at least appear to be – more profitable (often because their costs are not being fully counted or are partly offset by subsidies). As well, efficiency investments, such as weatherising private homes, often compete with pleasant idleness, which may be less profitable but also causes less bother.

The incentive structures in our societies seem not conducive to the efficiency revolution. Incentive structures historically were developed to encourage the ever fuller use of natural resources in the service of technological progress. Progress was new machines using energy in place of muscle power. Progress was an improved transport infrastructure and faster vehicles. To a large degree, progress was even the discovery and exploitation of new resources. According to unpublished papers circulating at the World Bank, direct and indirect subsidies totalling some $200 thousand million go into the energy sector alone.

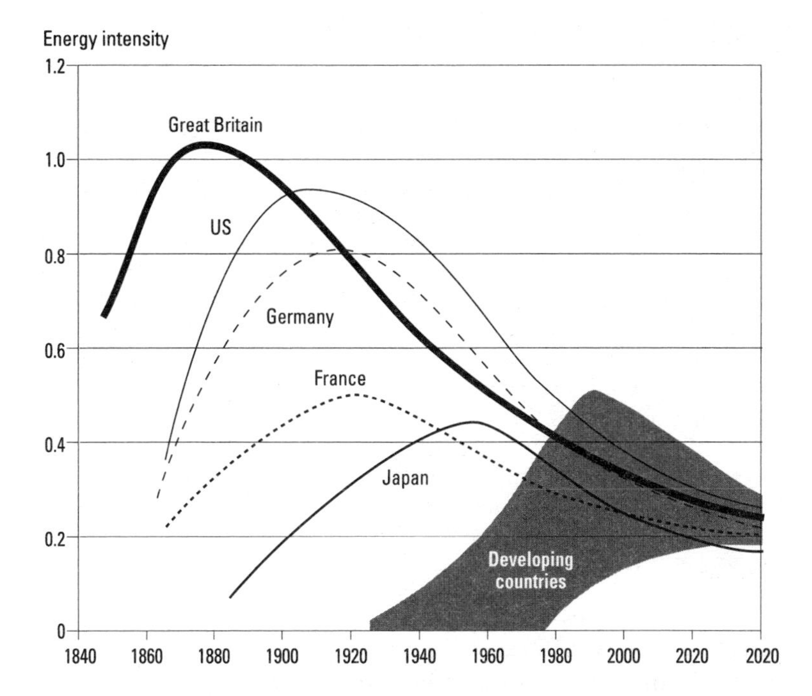

Figure 14 Energy intensity of industrialised countries since 1850. Two trends can be observed: (1) Energy intensity increases in the early stage of industrialisation, reaches a peak and declines in later stages (at a rate of roughly 1 per cent per annum); (2) the peaks become lower over the course of time.

Small wonder that throughout the period of industrialisation, up to and including our own day, resource productivity has had to fight an almost hopeless uphill battle. Capital typically has found and will continue to find it much more profitable to invest in the expansion of resource use (and to increase labour productivity). Capital typically finds returns on investments in resource use in the vicinity of 15 per cent or above per annum. This high rate cannot easily be reached by investing in many kinds of resource efficiency (though there are some striking counterexamples).

To be sure, there was some progress in resource efficiency over the decades (see Figure 14). Energy intensity, after reaching a peak for each country at a certain stage of its development along traditional patterns, is gradually receding, at a typical rate of 1 per cent annually. But, with very few exceptions observed between 1975 and 1985, this has not been enough to reduce the total energy consumption by any of the maturely industrialised countries. In any case, the trend came as a byproduct of technological progress and also as a result of an emigration to cheap-labour countries of some of the most energy-intensive industries.

If we assume, despite the geographical movements of energy-intensive industries, that the improvement in resource efficiency in OECD countries by 1 per cent per year could be extended to the whole world, the race would still be lost. This is because we observe and even welcome economic growth rates far above 1 per cent per year, so that energy consumption would still rise exponentially under this assumption. If global economic growth averages 3 per cent annually, resource productivity would have to increase at an annual rate above 3 per cent if we were to win the race. Given the proven potential in many sectors for productivity gains above 300 per cent (which is the percentage meaning of a factor of four), annual increases of 4–5 per cent would not seem an unrealistic hope, and in various times and places have already been achieved, even exceeded, as speeds of improving (say) energy productivity.

For this hope to materialise, however, it will be necessary to make resource productivity a highly competitive option for investments in the international capital markets as they are, *i.e.*, with their prevailing expectation of ROIs above 15 per cent annually.

In this part of the book we explore policy options for achieving that goal.

Chapter 4
If Markets Create the Problem, Can They Also Provide Much of the Answer?

4.1 Can Market Jujitsu Flip the Forces of Unsustainability?

Sir Winston Churchill said that democracy is the worst system of government – except for all the rest. Similarly, markets are the worst way to implement something profitable – except for all the other ways. Markets, like democracy, require ceaseless effort by their participants to keep them working properly and to prevent them from being subverted, distorted or hijacked by those who wish them to work improperly. But when they do work properly, they are phenomenally successful.

Markets have been so successful that they have often been the vehicle of runaway and indiscriminate growth that creates global problems. Markets are very good – all too good – at what they do. The modern market economy harnesses such potent motives as greed and envy; in fact, as Lewis Mumford caustically noted, all of the seven deadly sins except sloth. (Perhaps he was overlooking the entertainment industry.) But so effective is the profit motive that perhaps markets headed for unsustainability can best be redirected through creative use of market forces themselves, to harness their ingenuity, rapid feedback and diverse, dispersed, resourceful, highly motivated actors. This is, after all, the strategy of threatened biological systems that evolve new feedbacks and defences against a threat, often by turning the threat's own strengths against it – a common ploy of the immune system.

This chapter will suggest how markets can often provide much of the solution to their own excesses simply by steering their own immense forces, through carefully applied leverage, in more constructive directions. That approach is akin to the principle of jujitsu, which an American dictionary says literally means 'soft art' or 'pliant art', and defines as 'a Japanese system of wrestling in which the strength and weight of an opponent are used against him by means of anatomical knowledge and the principle of leverage'.

The idea that much of the answer to unsustainable market activity is sustainable market activity may offend both those on the right who don't see why what they're doing is unsustainable, and those on the left who think markets and profits can't be used for good ends. If so, that may be the price of pragmatism. For abundant experience now emerging in diverse societies suggests a bonanza of market-based institutional innovations no less important than the technological innovations just described. Ways are now known and proven to make markets in avoided depletion and abated pollution; to maximise competition in saving resources; to enable buyers and sellers of efficient techniques to meet and do business, so that a utility's cost of paying a sulphur tax can be directly converted into profit from selling a more efficient lamp or motor; even to represent future generations, through special financial instruments, in the investment decisions being made in their name. Perhaps the trouble with eco-capitalism is not that it has been tried and found wanting, but that it has not yet really been tried.

The basic principles of eco-capitalism – of saving the earth for fun and profit through advanced resource efficiency – are arrestingly simple. They include:

- Best buys first (find the cheapest ways to do the job, then buy them).
- Invest in saving resources wherever that's cheaper than extracting them.
- Make markets in saved resources.
- Use prices that tell the truth. Anything else is simply self-deception.
- Foster and monetise vibrant competition between all options on a level playing field.
- Reward the behaviour you want, not its opposite.
- Tax the undesirable, not the desirable.

- ❑ Scrap inefficient devices prematurely and replace them with efficient ones.

The examples in Chapters 4 to 7 show an extraordinarily rich tapestry of both proven and speculative-but-ready-to-try techniques. To be sure, they are as subject as any other market method to the inherent limitations of markets; but within those limits they are creative, vital and successful not only in supplementing and streamlining but often in supplanting prescriptive regulation.

Progress in saving energy in the US offers an encouraging example of how millions of small but widely accessible market actions can add up to very large gains in national resource productivity. Over the past 17 years, Americans got over four times as much new energy from savings as from all net increases in supply, and of those increases in supply, about a third came from renewable sources. This is especially remarkable because, throughout nearly all of that period, the federal government favoured expansion of large-scale, nonrenewable energy supplies such as coal and nuclear power and actively discouraged energy efficiency and renewables. But the market nonetheless preferred the best buys.

Already, the average American housing stock has become more thermally efficient than the German or Japanese, and many new American cars are more efficient than their German or Japanese counterparts. Competitive gaps in national energy productivity are gradually closing. And America's energy bills have already fallen by more than $160 thousand million per year (partly, of course, due to falling world market prices induced largely by slack demand caused by more efficient use). The millions of small improvements made by individuals, firms and communities – with insulation, caulk guns, duct tape, plugged steam leaks, slightly better cars – now provide two-fifths more energy each year than the entire domestic oil industry, which took a century to build. Within a few more years, energy efficiency will become America's largest source of energy services, surpassing all oil, domestic and imported. And nearly all of this occurred through everyday market decisions. Appliance and building standards, and utility programmes to help customers choose the best buys first, certainly helped, but mainly people just started becoming more aware of the nearly $2,000 a year each of them was paying for energy, and started demanding, getting and applying better information about alternatives.

Largely through better market choices, Americans have already saved an enormous amount of energy, contributing signally to low world oil prices for everyone. But Americans are still wasting more than $300 thousand million worth of energy per year – more than their government's entire budget deficit! Since energy-saving techniques and new ways to combine and apply them have improved even faster than the savings already achieved, the potential for saving still more energy is far greater and cheaper now than in 1973. And a broadly comparable potential, differing much in detail but little in broad outline, exists in every other country. To understand why it hasn't all been captured already, we need to inquire into the differences between theoretical markets and actual markets.

4.2 THE IMPERFECT MARKET

Factor Four's 50 examples, just presented, show that in many cases saving resources could cost less than buying and using them. This reduction in cost represents a theoretical profit opportunity. Chapters 4 to 6 describe some new and exciting ways to make it more likely that private entrepreneurs will realise that profit and hence gain a strong incentive to make resource efficiency the universal practice. Chapter 7 adds a tax reform proposal to speed it all up and further increase the business opportunities.

In an astonishing irony, one of the greatest obstacles to realising this potential – of creating huge new industries in resource efficiency – has been the mind-set of some of the free market's staunchest proponents. This is the naïve belief, particularly common among many free-market economists (and the ideologues they instruct), that existing markets are so close to perfection that any shortfall from the ideal is hardly worth examining, let alone taking much trouble over. In this view, if people live in draughty houses, it's because, after careful consideration, they have concluded that draughtproofing is not worthwhile. If a factory pollutes a river, it's because the factory's output benefits society more, and abating the pollution would cost more than any harm from the pollution. Or, in short, as Alexander Pope put it, 'Whatever is, is right.' After all, we live in a market economy driven by consumer preference as expressed through purchasing decisions; and if people wanted to buy something different from the society they've got, surely they'd already have done so.

This fatalistic dogma conveniently relieves us of the need to take any responsibility to change what's wrong, or even to acknowledge that anything is wrong. After all, saith the Market Doctrine, if a thing is worth doing, the market has already done it, and conversely, if the market has not done it, then it must not be worth doing. (This is reminiscent of the impregnable circularity with which economists define 'utility': people buy things because those things have utility for them, and the way we know the things have utility is that people buy them.) The assumption that the way things already are is indistinguishable from an economic optimum is an unexamined article of faith more than a reasoned conclusion. But it dominates public discourse and public policy, and it has consequences. This simple, and simply false, belief causes more than a million million dollars' worth of resources to be wasted every year.

Economists – of the sort who lie awake at night worrying about whether what works in practice can possibly work in theory – place the burden of proof on others to show that existing arrangements are not optimal and that market failures do exist. After all, the preconditions for a perfect market – perfect information about the future, perfectly accurate price signals, perfect competition, no monopoly or monopsony, no unemployment or underemployment of any resource, no transaction cost, no subsidy and so forth – are so elegant in their glittering austerity that it must be painful to descend to a world less perfect. But gentle ways can be found to remind economists of the distance between their theoretical world and the actual world the rest of us inhabit – the best way being to remind them that they live here, too.

One way, for example, is to ask how much electricity and money it took to run the economist's refrigerator last year. (Correct, indeed any, answers to this question are rare. Inability to provide an answer indicates market failure.) One can then ask if the economist knew that a model far (say, twofold) more efficient is on the market (generally a safe bet) with essentially the same price and features. It will often turn out that the economist, like most people, paid very little attention to energy efficiency when buying the refrigerator, or that someone else altogether (such as a landlord), who does not pay the bill to run the refrigerator, bought it. Usually, one will find that the economist's information about the refrigerator market is not merely imperfect, but virtually nonexistent. At some point the economist will object, 'I'm too busy living to take the trouble to find out all that.' Precisely: perfect information about available options, available without charge and digestible without transaction cost, does not exist as market theory requires. The same game can then be repeated with lamps, cars, computers or practically any other device the economist uses. In every case, the economist will typically be found to be making economically inefficient decisions because of poor information, the transaction cost of changing, lack of fair access to capital or other well-known barriers.

Transaction cost will deserve special attention. In its broadest sense it embraces the cost of changing laws, standards and norms; the costs of redesigning mass products, their assembly lines and their distribution logistics; the cost of writing off sunk capital; the cost of changing infrastructures, civilised customs (including values), habits of thought and education; the cost of overcoming ignorance on the part of consumers, producers, maintenance staff; the cost of removing bureaucrats who define their jobs in terms of dinosaur procedures; the cost of creating new jobs for workers now employed in the mega-inefficiency machine; and so on.

At some point, even the most theoretically devout economist will usually start to get the idea that market failures do exist and are actually experienced – even by economists. The economist may ask, 'Why should I insulate my roof? I rent my house, and although my heating bills would go down, the landlord would own the asset I was paying for. But I haven't been able to get the landlord to insulate the roof, because I pay the energy bills.' Just so: the classic 'split incentive' between those who pay and those who benefit. Or the economist may admit that when she invests in saving energy in her own home or business, she wants, like most of the rest of us, to get her money back in just a year or two – about ten times faster than energy companies want to get *their* money back from the power plants *they* invest in. This 'payback gap', requiring about tenfold better financial performance from saving than from producing energy, is equivalent to about a tenfold distortion in the energy price; it makes us buy too much energy and too little efficiency.

The reality and importance of market failures boil down to a simple story. Once upon a time, an elderly economist was taking his mannerly little granddaughter for a walk when she spied a £20 note lying in the street. When she asked, 'Please, Grandpa, may I pick that up?' he replied, 'Don't bother, my dear; if it were real, someone would have picked it up already.'

The economist's disbelief in windfalls is an empirical proposition subject to experimental test. Theoretical economists usually suppose that large, well-informed businesses have little opportunity left to save energy or other resources in a way that also saves money; any such opportunities, they assume, would long ago have been found and exploited by profit-maximising managers. But is this how the world really works? Scarcely. Consider the experience of Ken Nelson described in detail in Chapter 1, p 65.

It is a pity so few market economists have ever met anyone like Ken Nelson. Most economists will be hard pressed to believe the Dow example (or many more like it). It is hardly conceivable to them that such huge and juicy savings would have lain about untapped for all these decades, let alone that exploiting them should turn up even bigger and juicier ones all the time. This faith that what's worth doing has already been largely done is unfortunately not just an intellectual error; it has the disastrous practical consequence of preventing people from seeing what can be done.

During the Reagan/Bush era, a distinguished professor of economics at Yale University, William Nordhaus (1990), published a famous calculation supposedly proving that for the US to achieve the stabilisation of CO_2 emissions set by an international negotiating group in Toronto, and considered by most climatologists a modest first step on the path to stabilising the earth's climate, would depress the GDP (or, as national headlines trumpeted, would 'cost') about $200 thousand million per year. This astronomical 'cost' of even a preliminary climate-stabilising action got stuck in the head of the president's chief of staff, John Sununu, and paralysed policy on this issue for the rest of the Reagan and Bush administrations.

Dr Nordhaus's method of calculating the bill was simple:

- First he assumed that more efficient use of energy must not be cost-effective at today's prices, because if it were, people would have bought it already. Any market distortions were assumed to be immaterial, and no £20 notes were presumed waiting to be found. The existence of a huge empirical literature from people who actually sell energy efficiency, and who spend their days battling against a formidable array of all too real market failures, was ignored.
- Next, Nordhaus assumed that the only way to induce people to buy more energy efficiency would be to raise the price of energy, and that this could be done only through taxation. Since market failures had been assumed not to exist to any material degree, correcting those failures while leaving energy prices the same obviously wasn't an option.
- Next, he assumed that the revenues raised from the energy taxation would be rebated to taxpayers, so they could buy whatever they wanted, rather than invested. (This assumption makes GDP go down; investing the revenue makes GDP go up.)

- Next, he looked up historic studies of how much less energy people buy when its price rises. (This so-called price elasticity of demand is a shorthand way of summarising how people used to make millions of disparate decisions under conditions that no longer exist and that it is often a goal of energy policy to change as much as possible.)
- Finally, he turned the crank on a computer model to see how much energy taxes would have to be raised in order to depress energy use by an amount corresponding to the Toronto CO_2 targets, and how much that level of taxation would decrease total economic activity. The result: about $200 thousand million a year.

Actually, he may have got the number about right but the sign wrong. Meeting the Toronto CO_2 target would not *cost* but could in principle *save* the US about $200 thousand million a year, because saving fuel could cost less than burning it. Puzzled by this potential $400 thousand-million-a-year discrepancy, one of us went to Nordhaus's presentation of his thesis at a major scientific conference and asked, in the discussion period, why he hadn't used the vast empirical literature on how much it actually costs to save energy, as measured and documented by thousands of utilities and industries that actually do it every day. His reply: 'I just used an assumption from economic theory. That's an interesting hypothesis you have, Mr Lovins, that a lot of energy-saving measures are cost-effective at today's prices but aren't being bought because of some sort of market failure. Of course, if you used that assumption instead of mine, then you'd get a very different answer.' But he refused to take responsibility for how his theoretical assumption had stymied global efforts to approach climatic questions in a least-cost, best-buys-first sequence. In fact, he was so fond of theory that he didn't want to consider the facts, and didn't seem to know the difference between the two.

4.3 MARKET THEORY VERSUS PRACTICE

At the Santa Fe Institute, where scientists explore the frontiers of mathematically chaotic systems, there is a saying that 'In theory, theory and practice are the same, but in practice, they're not.' Markets are like that too, and it's very important to understand the difference.

Market theory, for example, would suggest that a device that, say, uses energy efficiently will cost more than an otherwise equivalent device that doesn't. After all, to save the energy, the device must substitute more of something else, such as materials or brains, and that substitution incurs a cost. Or does it? Chapters 1 to 3 gave several examples showing that smarter combinations of technologies, or properly counting multiple benefits ('joint outputs'), could often make big savings cheaper than small savings – contrary to the theoretical assumption that the more we save, the more each increment of savings must cost. But even with single, simple devices with no joint outputs, empirical data show that efficiency needn't cost more. Unlike those theoretical economists who prefer theory to facts, let's go look at some facts, following electric car pioneer Alec Brooks's motto: 'In God we trust; all others bring data.' The data may surprise us.

- Figure 15 shows the market price of all household refrigerators on the Swedish market compared with their electrical consumption, both per litre of interior space. Not only is efficiency unrelated to price, but the most efficient model (the LER200, made by the Danish firm Gram) costs less per litre than some other models that use six times as much electricity per litre! Similar results can be shown for most major household appliances, from washing machines to stoves and from freezers to televisions.
- The same absence of correlation has been established between the trade prices of 5–20-ton (a measure of capacity) mass-produced rooftop air-conditioning units, often used in commercial buildings, and their energy efficiency ratings. Similar data exist for the trade price of three different sizes of the most common kind of industrial electric motor, and for large industrial pumps. These data are far from comprehensive, but they're certainly suggestive that resource efficiency doesn't always cost more to achieve. In fact, they suggest that, on this basic point, economic theory is so suspect that it should be rigorously tested against real data from a wide range of fields. Nobody has yet done so.

How about another standard assumption of economics – that people use more of a service or a good the less they have to pay for it? Willett Kempton et al (1992) have looked at the behaviour of people using air-conditioners when they believed that both the air-

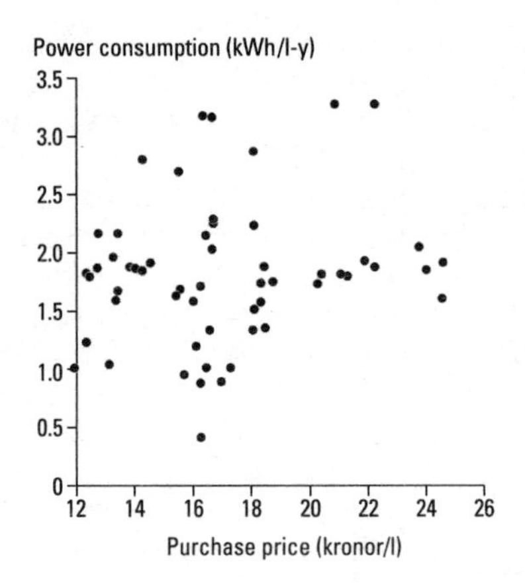

Figure 15 Energy efficiency doesn't always cost more. No correlation can be found between energy efficiency and the prices of refrigerators in Sweden. The findings stem from work carried out at the University of Lund.

conditioners and the electricity used to run them were free. Yet almost nobody ran the air conditioners all the time (or whenever they felt hot); on the contrary, many didn't run them at all, and for those who did, the patterns of operation had very little to do with comfort. The reasons given for turning air conditioning on or off related instead to the humming noise which people preferred to the noise from the road, or with incorrect theories about the working of thermostats, or to complex belief systems about health and physiology and so on.

There is no reason to believe that the economists' assumptions about *Homo oeconomicus* are any better founded in empirical reality than those of engineers or psychologists or sociologists. Any one of such disciplines' views tells a true story, to be sure, and each has its own sphere of validity. But each is a kind of mental map, and the map is not the territory.

That is, in the real world full of complex people, both the engineering and the economic models of how people behave are not only incomplete but often seriously misleading. The concept that consciously or unconsciously underlies economists' pronouncements of how 'rational consumers' will behave, and hence how their choices will depend on prices, policies and other conditions in society, is only modestly more compelling and complete than earlier methods involving, say, inspecting the entrails of slaughtered fowl.

The only difference is that, today, economics is the state religion to which policies are aligned, hymns sung and goats sacrificed daily, while the old Roman methods of divination are out of fashion.

Our point here in casting (nay, heaping) doubt on the universal validity of economic theory is not to suggest that it is useless, but to warn that, like any theory, it must be used with caution, discrimination and healthy scepticism. The next two chapters suggest many ways to apply economic methods to overcome market failures and achieve worthwhile social outcomes. But we must always stay warily alert to something every good marketer knows: that human choices depend on far more than price, and must be influenced by many means wholly unrelated to price. People are far more complicated than that.

The Litmus Test

Economics, for all its gaps and uncertainties, remains an important intellectual construct with profound practical consequences. The difference between a market economy, however imperfect, and a centrally planned economy – like the old Soviet command economy, which was essentially a giant machine for eating resources – remains real and crucial. But it does not follow that anyone pointing out flaws in the actual operation of market economies, or calling for sound market principles to be seriously applied, is thereby favouring command-and-control or central-planning principles. On the contrary, Chapters 6 and 7 will next show how even very imperfect markets can become remarkably effective tools for harnessing human motivation and potential, thus reducing any need to substitute bureaucratic for private preferences.

But are bureaucracies ready to give up control? The rest of the 1990s will offer a fascinating test of political *glasnost* in societies that consider themselves market-oriented. Will they really wean themselves off corporate socialism and adhere instead to the market principles they preach? Will the US, for example, eliminate its more than $30 thousand million a year in federal subsidies for energy supply – nearly all going to the least economical options that remain the least successful in the marketplace? (Until 1986, those subsidies exceeded $50 thousand million per year, but they're even more lopsidedly distorting today than they were then.) Will Germany's three big generating utilities, accustomed to a cozy cartel-like arrangement, start publishing basic data on how much electricity is made by whom and sold to whom at what price, opening up their grid to private generators, and letting everyone compete fairly? Will Électricité de France come clean about its (and its nuclear suppliers') enormous but still secret subsidies? Will power companies worldwide start allowing real competition in who makes and who saves electricity?

When Ronald Reagan was elected president, two of us published a *Washington Post* commentary titled 'Reagan's Energy Policy: Conservative or Ultraliberal?' It suggested a number of litmus tests for determining whether the new administration's energy policy was true-blue (conservative) or a suspicious shade of red (socialist). Would it desubsidise the energy sector and let all options compete on economic merit? Would it level the playing field so that new market entrants could compete fairly with older and more established ones? Would it subject nuclear power, synfuels and other bureaucratically mandated options to the same discipline it proposed for solar and efficiency options, or would it instead foster central planning for the options it favoured and free enterprise for the rest? In the event, alas, the litmus turned a shocking pink: a government professing to follow market principles turned out to be a band of corporate lemon socialists, dedicated mainly to rescuing their favourites that were dying of an incurable attack of market forces. (The recent 'sale' of British Energy, practically paying 'buyers' to haul the carcass away, illustrated the persistence of such attitudes.) The same political choices face us today – with the added dimension that efficiency and renewables are by now so much more highly evolved that even their opponents must concede they represent a real threat to the established order.

All economies are mixed economies – part free-market, part instructed or guided or influenced by government policy to produce desired outcomes despite market forces. The main difference today is between countries that admit this and countries that pretend they're not distorting market outcomes. But in any society, markets present vastly greater creative opportunities for profitably promoting resource efficiency than have yet been harnessed. The next two chapters suggest ways to make eco-capitalism a vibrant reality – to help markets work as they are supposed to. But even then, one note of caution is required: markets are only tools, not a religion; means, not ends. They can be used much better to do many important things, but they cannot do everything, and it is dangerous to suppose, in particular, that markets can replace ethics or politics. In Chapter 14 we shall return to the many differences between the way market economics is conventionally practised and the way its founders rightly proposed – and between the purposes of markets and of human beings.

Chapter 5 Buying and Selling Efficiency

5.1 Least-Cost Planning

How do we get the efficiency revolution going in a market economy? How can efficiency be bought? How can it be sold at a profit? Has there ever been a company that attempted to sell less of what it produced? In fact, there are such companies, indeed a whole industry: American electric utilities, the largest single sector of the US economy, with an asset value over $500 thousand million and an annual cashflow of nearly $200 thousand million. This sector has been undergoing a systematic transition from worst to best buys first, and from central planning toward market mechanisms.

First a little history. Monopolistic abuses early in the 20th century led every one of the United States (except Nebraska, whose utilities are all publicly owned) to evolve a complex system of largely private utilities, three-quarters owned by market investors – regulated by state commissions that set electricity prices and must usually approve major investments. The principle behind this regulated-monopoly approach was that distributing electricity was a 'natural monopoly' – it didn't make sense to have more than one set of wires running down the street.* And, in a matter so 'peculiarly affected with the public interest' (to use the words of the famous jurist Learned Hand), a degree of politically accountable oversight was vital to ensure a universal, fair, nondiscriminatory, reliable, healthful and safe supply of electricity.

* About 17 US cities and towns, some of them quite sizeable, nonetheless do exactly this, having both public and private electric distribution utilities competing with each other and reportedly tending to lower prices. Economic theorists have paid little attention to these counterexamples.

The states therefore evolved elected or appointed public utility commissions (PUCs for short, though their actual names varied), charged by law with upholding these social interests while ensuring continued viability of the utility by providing investors with a fair, just and reasonable return. Until the mid-1970s, being a PUC member did not seem very important: the utility would propose when and where to build its next power station, and after fairly desultory discussion, the PUC would rubber-stamp the application. Prices drifted downward as power stations became bigger, cheaper per kilowatt, and more efficient and reliable, so PUCs had the delightful task of allocating the savings to various kinds of customers as ever-lower tariffs.

But in the 1970s, this comfortable world imploded. The bigger power stations started to become more costly and less reliable. Hugely ambitious commitments to unproven nuclear technology in projects of unprecedented size and quantity led utilities into uncharted territory that soon overwhelmed their technical and managerial capacities. The Arab oil embargo in 1973, followed by the more severe second oil shocks from the 1979 Iranian revolution, dramatically boosted inflation, interest rates and fuel prices. Many utilities skidded towards bankruptcy – a fate that in the end did claim several of them. Dramatically cheaper alternatives that substituted more efficient use, or other fuels, for costly electricity posed market threats – and opportunities – to the beleaguered utilities. And in 1978, Congress opened electric generation to competition by all bidders by requiring utilities to transport ('wheel') power privately generated by others, and to pay a fair price for it. Every assumption of the utilities' comfortable world was turned upside-down in the 1970s.

From this searing intimation of economic mortality emerged new operating principles, clear in their economic purpose and striking in their logical simplicity:

- Customers wanted not kilowatt-hours per se, but hot showers and cold beer, comfort and illumination, torque and electrolysis – the 'end-use services' that the energy provided.
- Customers wanted to get those services reliably, conveniently and in the cheapest way possible.
- Kilowatt-hours of electricity therefore had to compete with all other ways to provide the same service: through increased electrical productivity (using fewer kilowatt-hours and more brains to do the same task with smarter technology), or by substituting other fuels, or by substituting renewable energy.

- As customers figured out that 'negawatts' – saved electricity – were cheaper than megawatts, they would naturally want to buy less electricity and more efficiency in using it.
- The only question was who would sell them the efficiency.
- It is a sound business strategy to sell customers what they want, before someone else does.
- Utilities' only choice, therefore, is between participation in the 'negawatt' revolution and obsolescence.
- PUCs could help utilities develop ways to compare all methods for a customer to do a given task, determine which is cheapest and sell the customer that cheapest method first.

Best Buys First

From this last idea emerged the process of 'least-cost planning', in which utilities could compare all options for providing a given end-use service, choose the cheapest, and buy it or help the customer buy it. 'Cheapest' could reflect full social costs or could be constrained to narrowly defined private internal costs; either way, the basic principle was to choose – and then buy – the best buys first.

Over the years, most PUCs in the US, and many analogous bodies abroad, adopted this approach in varying degrees. By 1992, 'integrated resource planning', the official name for essentially the same concept, was required of every state by federal law, and only rarely could unregenerate utilities get away with comparing only different kinds of new power plants rather than a wide range of ways to make electricity, distribute it better, or displace it with cheaper substitutes like efficient end-use or alternative fuels.

The least-cost principle of asking What's the job? What's the best tool for the job? began to spread beyond the electricity industry, first to natural gas (regulated at the retail level by the same PUCs), then to water (often also under PUC regulation), then to other spheres such as transportation. A 1991 law called ISTEA (the Intermodal Surface Transportation Efficiency Act) called for a wide range of transportation alternatives, including demand reductions, to be considered whenever a new highway was planned. Amendments to the Clean Air and Clean Water Acts had a similar effect, calling for demand-side, efficiency-based alternatives to receive fair consideration alongside supply-side options.

Just considering cheaper alternatives, however, turned out to be only one of three basic ingredients of an effective way to provide more and better services with less money and trouble. The other two proved more subtle. They involved, first, reforming utility regulation, and second, developing increasingly market-oriented ways to translate least-cost choices from a paper exercise into actual purchases and deployment. We consider these two parallel revolutions next.

5.2 Utility Regulatory Reform

The standard way for American PUCs to set electricity prices – a practice dating back for the best part of a century and eventually adopted nearly worldwide – started by determining, from various market and political judgements, what return on investors' employed capital would be fair and would continue to attract such capital as the enterprise required. Multiplying this target 'rate of return' times the 'rate base' of used and useful capital (and counting its amortisation) would then yield the amount of revenue the utility would need, say, next year for a return on and of that capital. To this would be added expenses reasonably incurred for operations. This total 'revenue requirement' would then be allocated to the various classes of customers (residential customers, large industrial users and so on) in a fashion deemed to represent fairly their respective fractions of the total cost of running the system.

But here entered a complication whose perverse implications were not really understood until the mid-1980s. To convert revenue requirements into actual tariffs (how much each kind of customer would pay for each kilowatt-hour), the PUC would have to assume how many kilowatt-hours each customer class would buy next year. If the utility then actually sold more than that number of kilowatt-hours at the price thereby determined, it could earn more profit, whereas if it sold fewer kilowatt-hours, its profits would fall. This in turn created an incentive to 'game' the forecast: the utility would try to understate its expected sales, while its customers, represented in the PUC hearings, would try to overstate sales. This caused long and unconstructive debates between their costly expert witnesses and lawyers. Moreover, the utility's profits now depended on things it could not control, such as weather and business conditions; and there is no public-policy reason to make the utility's profits depend on things it cannot control.

A further complication – and increased uncertainty for customers – arose from the practice, nearly universal by the 1980s, of 'fuel-adjustment clauses'. These clauses automatically passed through changes (usually, in the aftermath of the 1973 and 1979 oil shocks, increases) in fuel prices into changes in electricity tariffs in order to save the PUC from having to hold new hearings every time volatile world oil prices, and prices of other fuels, changed. This turned out to exacerbate the problems of traditional price formation.

Decoupling Profits from Sales Volumes

Starting around 1980 in California, some states 'decoupled' utilities' profits from how much electricity they sold. PUCs changed the rules so that the utility wouldn't profit by selling more electricity than expected (the excess profit would simply go into a balancing account, not to the utility), and conversely, wouldn't suffer by selling less electricity than expected (money would be returned from the balancing account to the utility to make it whole).* This eliminated the incentive to game the forecast and the dependence of profits on weather and other uncontrollable factors. Utilities liked it because it reduced their financial risks and helped them plan more soundly. In a given year, tariffs might fluctuate significantly, but over time, the net effect on tariffs was virtually zero – one-quarter of 1 per cent during the first decade in California, for example.

Sharing the Savings

Then emerged a further problem: for a variety of accounting reasons, some of them rather subtle, utilities weren't getting nearly as much profit from saving electricity as they could by selling more of it.** This perverse incentive led them to build more power stations and to expand the grid even when that wasn't the best buy for the customers. So some PUCs introduced the second part of basic reform by letting the utility keep as extra profit part of whatever it saved its customers. Sharing the savings gave both parties an incentive actually to achieve them.

Utilities that weren't rewarded for selling more electricity, weren't penalised for selling less and were well rewarded for cutting their customers' bills then did something so predictable for market theory that almost nobody expected how powerful it would prove: they started investing strongly in end-use efficiency in order to cut

---

* Some states instead have a 'lost revenue adjustment mechanism' that is meant to produce the same result but actually doesn't. This approach is becoming less popular, partly because it's far more complex.

** Some early ways to deal with this, such as adding efficiency investments to the rate base (so utilities could earn a return on them just as they did for power stations), or giving them a higher rate of return for efficiency investments, turned out not to make the comparison fully symmetrical, or to introduce unexpected new problems. Some of these inadequate methods are nonetheless still in use.

customers' bills by saving electricity more cheaply than it could be produced. The 'negawatt revolution' was well underway. And in late 1989, when the National Association of Regulatory Utility Commissioners, in a rare instance of unanimity, endorsed the concept that the best buy for the customer should be the most profitable investment for the utility, regulation that emulated market principles by rewarding economically efficient behaviour became – at least in principle – the accepted norm.

THE PROOF OF THE PUDDING

Consider, for example, Pacific Gas and Electric Company – the largest investor-owned utility in the US, serving most of Northern California. Around 1980, PG&E was planning to build some 10–20 power stations; it envisaged a nuclear station every few miles along the state's entire coastline. But by 1992, PG&E was planning to build no more power stations, and in 1993, it permanently dissolved its entire engineering and construction division. Instead, as its 1992 *Annual Report* announced, it planned to get at least three-quarters of its new power needs in the 1990s from more efficient use by its customers, and the rest from the second-best buy – privately bid renewable sources. If it needed more, which it didn't expect to do, the third-best buy would be advanced natural-gas-fired power stations (combined-cycle and steam-injected gas turbines). Coal and nuclear plants, once considered the only practical options, were now viewed as so relatively costly that they'd never again be considered.

What caused that revolutionary change in this giant utility's plans and practices? Largely that its profits no longer depended, in either direction, on how much energy it sold, and that its shareholders got to keep as extra profit 15 per cent* of whatever savings the company achieved for its customers – whether by helping them use energy more efficiently, buying cheaper fuel, running its facilities more efficiently and reliably, or making any other operational improvement. In 1992, therefore, PG&E spent more than $170 million helping its customers to save electricity – the largest such programme in the world. That single year's investments yielded net benefits worth some $300–400 million (expressed as a lump sum in 1992). Of that created wealth, the customers got 85 per cent as lower bills, while the shareholders got the rest – more than $40 million – as higher dividends. The company also recovered the costs

* This share was later raised to 30 per cent, albeit of a smaller number, as the 'avoided cost' that the utility could save by helping customers save electricity became much lower in a more competitive market awash with cheap natural gas.

of the efficiency investments from all its customers, just as they would otherwise have paid for a power station (only cheaper) – a fair disposition of the costs, since efficiency was the cheapest resource for the utility system to acquire.

You can readily imagine that if you run a department that adds more than $40 million to the bottom line (the company's second-biggest source of profit), at no cost and no risk to the company, the chief executive is likely to call you up every week to ask, 'Is there anything you need?' – and all the smartest people in the company will want to come work in your department for their career advancement. That is, rewarding the behaviour we want (lower bills) rather than the behaviour we don't want (selling more electricity – the previous basis of utilities' profits) can quickly change not only what a utility buys, but its whole mission and corporate culture.

In 1993, the California PUC published confirmation that during 1990–93 alone the efficiency efforts of the utilities it regulates had saved California customers nearly $2 thousand million more than they'd cost. In 1994, the PUC also released the results of an exhaustive review of hundreds of field studies confirming that the efficiency programmes had in fact saved almost exactly the amount of energy predicted, and had done so at far lower cost than producing the same energy. The art and science of measuring energy savings are now so highly developed that these things can be rather easily and accurately determined at very reasonable cost.

On a more modest scale, attempts have been made in Europe to establish a least-cost regime for energy supplies. There is no equivalent in Europe to the PUCs, and in many cases municipalities can, through city-owned energy distributing companies, enforce the investments in cost-effective energy efficiency. Peter Hennicke, director of the Wuppertal Institute's energy division, has successfully cooperated with various such local utilities (Stadtwerke) in Germany to show them the way to energy efficiency. Figure 16 shows for the Stadtwerke Hannover how much energy can be saved cost-effectively, *i.e.* at costs below the typical price of additional power supplies.

The Retail-Wheeling Confusion

Another development in 1994 was less encouraging. On 20 April of that year, a small, secret cabal of California PUC commissioners and staff stunned the US utility and regulatory communities by proposing to scrap these highly successful practices in favour of an ideological agenda called 'retail wheeling'. Under this scheme, any utility customer could pretend to be buying electricity directly and

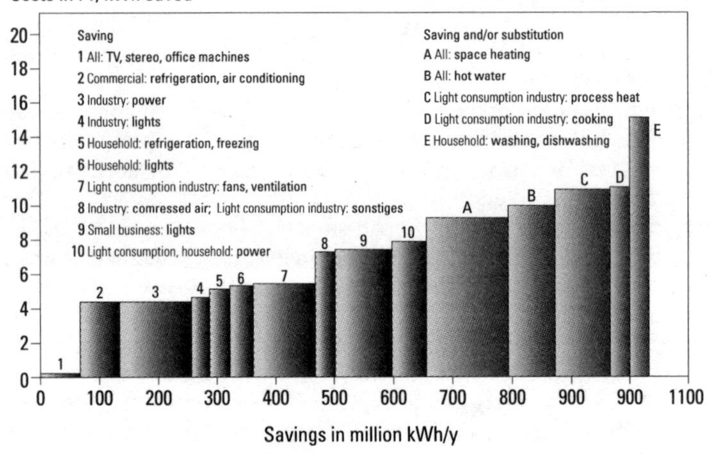

Figure 16 The planned Hannover 'negawatt plant'. Some 600 million kWh can be saved at no loss of service quality and at cost below 7 Pfennig per kilowatt-hour, i.e. profitably. Another 300–400 million kWh is available at below 10 Pfennig per kilowatt-hour. The plan is equivalent to the construction of a new 300-MW power station. (After Hennicke and Seifried, 1996, p 122)

competitively from any supplier, at prices they alone would determine.* This agenda had actually been advanced by a few large industrial customers who wanted to shift the costs of the most expensive (mainly nuclear) power stations to smaller and weaker customers, rather than fairly sharing all costs among all customers and rewarding utilities for reducing those costs. Cloaked in the superficially attractive language of competition and choice, retail wheeling is really about shifting costs, not reducing them. Indeed, virtually all of the savings it is claimed to offer would already have been captured by wholesale competition, where utilities or groups of utilities shop for the cheapest bulk electricity – a money-saving practice already required by federal law since 1992 and lately starting to be widely implemented.

There are striking similarities between retail wheeling and the draft European Third Party Access (TPA) Directive agreed in July 1996 but published in the *Official Journal* in 1997. It reflects the

* A B Lovins, Is There Life After the CPUC Order? address to the National Association of Regulatory Utility Commissioners National Conference on Integrated Resource Planning (Kalispell, May 1994), RMI Publication no. U94-17.

intention laid down in the Single European Act of 1987 to establish free markets for electricity, and will eventually lead to a dismantling of the regional power monopolies. Free market competition will lead to lower energy prices, and thus to reduced benefits from energy efficiency. But special clauses, already agreed in Denmark, can reserve a small percentage of the energy revenues to finance efficiency programmes or support renewables. Both efficiency and renewables would be greatly encouraged if the draft directive on integrated resource planning was adopted. However, the directive which was proposed by the Commission in September 1995 to introduce the benefits of least-cost planning has been stalled by Germany, largely due to the pressures of her energy-intensive industries.

Retail wheeling would also scrap the present practices of counting environmental costs in utilities' resource decisions (so those costs would no longer be counted at all) and of fostering a cleaner, safer, more diversified and more renewable portfolio of power sources. By throwing away the past century of utility regulatory practice, it would virtually eliminate safeguards that now ensure fairness, reliability, nondiscrimination and other important aspects of the public interest. And by resuming outmoded concepts that reward utilities for selling as much electricity as possible, retail wheeling would not merely stop rewarding investments in efficiency, but would positively penalise them – much as Britain mistakenly did by adopting misguided incentive structures during its electricity privatisation, thereby destroying what little efficiency industry it had.

The press has widely and wrongly reported that these changes have already been adopted in California and are inevitably sweeping across the US. In fact, however, they have not yet been adopted anywhere (except for a few tiny experiments unlikely to be widely imitated). Nor is this likely, because retail wheeling would require enormous changes in basic state and federal laws, and faces formidable practical and political challenges. But the perception that retail wheeling is happening or at least is inevitable – based on misunderstanding of the lessons of previous restructuring and deregulatory trends in other industries – has already created paralysing confusion in the US utility industry, setting back efficiency efforts by several years.

Many utility managers have a herd instinct, and have adopted the current fashion of supposing that, in a competitive environment, there will be no room for customer-service 'frills' such as efficiency services. (In fact, of course, once they buy bulk power at competitively levelled prices, there is no important way to differentiate their service from other distributors' service except by bundling electrons with more efficient use to cut customers' bills.) Moreover, investors, correctly scenting a massive confiscation of shareholder assets, wiped $50–100 thousand million off US utilities' share values.

Once cooler heads prevailed, however, from the 1994 retail wheeling panic has now begun to emerge a useful new understanding of how best to capture the savings available both from buying wholesale electricity competitively *and* from using it more efficiently. Efficiency savings dwarf competitive wholesale buying savings. Retail wheeling would sacrifice the former to the latter. But there is no reason we cannot have both – by fostering vibrant wholesale competition, buying power from our local utilities,* *and* rewarding them for cutting our bills rather than for selling us more electricity. That way, the interests of utility shareholders and customers will be identical, not opposite. Everyone will win as net savings from electric efficiency grow from the $5 thousand million a year already achieved in California to well over $100 thousand million a year nationwide. Costly old (mainly nuclear) plants will die in peace. Promises will be kept; and fairness, not greed, will govern a stable, clean, farsighted and economically efficient power system. In most of the US, though not yet elsewhere, such a farsighted electricity system is now starting to emerge (Lovins, 1996).

5.3 MAKING NEGAWATT MARKETS – AND BEYOND

Achieving this will require one more kind of innovation: continued improvement in how well and how cheaply the energy-saving technologies actually get delivered to the customers, and how far this is done, not by bureaucratically driven utility programmes but in the marketplace.

* A few states, including California, are promising that if customers really think they can get a better deal buying power from the wholesale 'pool' themselves than letting their utilities buy it for them as aggregators and purchasing agents, they are welcome to try—but they will not thereby be able to escape paying, through a wires charge (now also required by federal law), for their fare share of the old stations built to serve them, nor for 'public goods' costs such as health, safety, environment, R&D and cost-effective energy-efficiency investments. Though sometimes called 'direct access' or even 'retail wheeling' as a political sop to its advocates, this is a far cry from the retail wheeling originally sought by large industries seeking only to shift their costs to other customers.

From Information to Financing

In the 1970s, when US electric and gas utilities first started helping customers to save energy – because even then, that was cheaper than supplying more energy – they thought at first that information would be enough. Inform the customers of their choices, the theory went, and they would buy the right thing. Many information programmes were developed, some general and some specifically targeted to the technical needs of shopkeepers, architects, engineers, electricians, plumbers and other key actors.

But it turned out that information alone, without financing, accomplished little, largely because of the 'payback gap' that imposes roughly tenfold more stringent financial requirements on saving energy than on supplying it. (This presents the obvious business opportunity of arbitrage on the spread in discount rate between the consumer and producer – a spread on the order of a thousand percent, compared with the fractions of 1 per cent on which arbitrageurs normally expect to make money.)

For this reason, some utilities started to finance their customers' insulation, draughtproofing and other home improvements at low or zero interest. Then they discovered that it was often cheaper to give the weatherisation away than to service and subsidise the loan. (Many water and sewer authorities already give away high-performance showerheads and other inexpensive water-saving devices because it's the cheapest, most foolproof method of delivery and achieves savings worth vastly more than it costs.) Meanwhile, utilities invented the 'rebate' – helping to level the playing field so that ways to make and to save energy would be assessed on equal financial terms, by making a specified monetary payment to anyone who actually provided savings to the utility.

Rebates

At first, rebates were paid only to the purchasers of specific kinds of efficient equipment: so much for a refrigerator using no more than so many kilowatt-hours a year, so much for an electronic lighting ballast. But then utilities realised that savings are all interchangeable, and that customers may come up with methods of savings that the utility hadn't thought of, so rebates started to become 'generic': you save, we pay, we don't care how you do it. Payments of several hundred dollars per saved kilowatt, or its equivalent per saved kilowatt-hour, became commonplace. And then some utilities started to offer part of the rebate to the wholesaler, the retailer, the specifier, the installer – anyone who had to be paid something to get their attention and make the sale happen. By now, perhaps a quarter of US utility rebate programmes reward 'trade allies', not just the retail buyer of the equipment. Sometimes those rewards are not in cash but in kind, such as training or promotional programmes.

For example, PG&E found that, rather than paying 1.9 cents to save a kilowatt-hour by rebating buyers of energy-efficient refrigerators, it could achieve a bigger saving at a cost of only 0.6 cents per kilowatt-hour by giving a cash bonus of $50 to the retailer who *sells* an efficient refrigerator – but nothing for selling an inefficient one. The inefficient ones then quickly vanish from the shops; most likely they are shipped to areas without such an arrangement. (It is dangerous to be last in such a queue.) When a PG&E customer's refrigerator fails and the next shippable unit is urgently required, it will then be efficient, because that's all that's available.

Around 1991, Southern California Edison Company realised that its retail rebates would yield far larger savings if moved upstream to the manufacturer. For example (Figure 17), SCE was giving a $5 rebate for buying a compact fluorescent lamp in a retail shop, cutting its effective price from, say, $19 to $14 – still too high psychologically to crack the household market. But if instead the same $5 rebate went to the manufacturer for every lamp ultimately sold in SCE's territory (as verified by trade channels), the $9 manufacturing cost would be effectively cut to only $4. Once marked up by the wholesaler's and retailer's normal percentages, the retail price would be only $9. And then SCE said, 'Our promotion is causing far more lamps to be bought, we're doing a lot of promotion and production costs are dropping at these higher sales volumes, so the manufacturers should give us back some of their profit margin.' They did, dropping the net manufacturing cost, after rebate and discount, to only about $2.50. Marked up by the same traditional trade percentages, the retail price in the shop was now barely over $5.50. And everyone made more money, because the dealers gained far more on volume than they lost on the dollar mark-up from the lower base price.

You might at first suppose that rebates should be kept as low as possible to avoid throwing money away, especially if you suspect that a substantial number of 'free riders' will accept the rebate who would have achieved the saving even without it. But this seemingly commonsensical approach is not the best idea if in fact a high rebate, at least for a prescribed period, will achieve an even broader 'market transformation'.

That is what British Columbia Hydro did starting in 1988 to

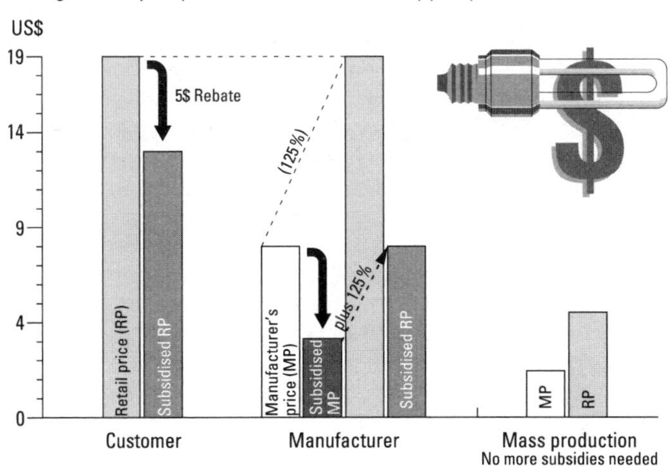

Figure 17 Southern California Edison discovered that subsidising efficiency equipment at the end consumer's level is not as cost-effective as subsidising it at the manufacturer's level. Left bars: A $5 subsidy to a $19 lamp at the consumer's end results in a price of $14. Centre bars: Spending $5 per lamp at the manufacturer's end cuts prices at the factory from $9 to $4. Adding a typical 125 per cent from factory to retailers results in a retail price of $9. This may result in dramatically increasing demand, allowing the manufacturer to mass-produce and decrease the price to $2.50, leading to a retail price of $5.50 – yet at higher profits for all.

transform the market for premium-efficiency industrial motors. Big motors used a large fraction of the Canadian province's total electricity, chiefly in huge mining and pulp-and-paper plants, but they were virtually all low-efficiency models, because premium models were simply not stocked: they required a special order and a long wait, intolerable to industrialists who needed to restore the failed motor to service overnight. But in only 3 years, B.C. Hydro raised the market share of premium-efficiency motors from 3 per cent to 60 per cent by offering rebates so generous that nobody could afford to ignore them. Now the market has completely reversed: it's the standard-efficiency motors that are special-order items incurring a long wait. And to make sure the dealers and end-users retain this new habit, Hydro is supporting legislated efficiency standards, then will ramp down the rebates as they become superfluous.

VARIATIONS ON THE REBATE THEME

More good ideas began to appear in the 1980s. Some utilities offered rebates not just for buying efficient new equipment but also for scrapping inefficient old equipment so that nobody would ever use it again: so much for your old refrigerator, with the few best ones to be given to low-income households (replacing their less efficient ones) and the rest to be dismantled, recovering the CFCs from refrigerant circuit and insulating foam and recycling the metal. This approach offers huge potential if extended beyond household appliances. To take a negative example, if it is not applied to industrial motors, then the inefficient old motor displaced by the efficient new one will enter the secondary market, and next week the utility will see it again reinstalled down the street (or, even worse, sold to a developing country whose overstressed power grid can least afford it). Part of the rebate for the efficient new motor needs to await a 'death certificate' for the old one.

Still another why-didn't-we-think-of-this-long-ago idea was paying rebates for better design – a topic so important we reserve it to Chapter 6, p 177. Some utilities now pay rebates for beating appliance or building standards – the more you improve on them, the bigger your rebate. This elicits new technologies, and the standard can be progressively raised until the cost-effective opportunities have been exhausted.

In an effort to jump-start the gradual improvement of the standards themselves, a utility consortium in 1993–94, building on a successful earlier experiment in Sweden, offered a 'Golden Carrot' – a big cash prize payable to the winner of a competition to bring a doubled-efficiency, CFC-free, no-compromises refrigerator to the US market. The prize, won by Whirlpool, will be paid per qualifying unit actually sold, so the utilities are at no risk: they pay only for the electrical savings they actually get. The losing manufacturers must also follow suit without being paid; otherwise they will lose market share and will also fail to comply with legal standards when, following the winner's example of what is possible, those standards are later tightened. And the winning manufacturer's risk for its R&D investment is greatly reduced by the knowledge of guaranteed payment for the first x units it actually sells – and by the participating utilities' offers of promotion and even outright purchase. (Public housing authorities, military procurement and other bulk buyers can often aggregate large guaranteed markets in this way to elicit an improved device from manufacturers.)

Yet another variant, still at an early stage of deployment, is for utilities to lease efficient equipment to customers, who can then pay for it over time exactly as they otherwise pay for a power station over time, only cheaper. For example, a few utilities have leased compact fluorescent lamps for, say, 15 pence per lamp per month, replacing them free (for a continuing lease fee) when they burn out. Some utilities are even considering leasing major items of efficient industrial equipment. This can be advantageous to both parties, and can even give rise to certain tax savings in some cases.

Moving to the Market

All these traditional ways to market negawatts – information, gifts and rebates – can be very effective in maximising the number of people who participate and how much energy each of them saves. But they lack an important further element: maximising *competition* in who saves and how so that we can continuously improve the cost and quality of the savings. To do that requires not just marketing negawatts but making markets *in* negawatts: making saved electricity (or other saved resources) into fungible commodities, subject to competitive bidding, arbitrage,* derivative instruments,** secondary markets,*** and all the other mechanisms that apply to copper, wheat, sowbellies and other traded commodities.

A basic feature of commodities is that they are bought and sold by competitive bid, at prices that change all the time depending on

* This means profiting from the difference in price between the same thing being sold in two different places at the same time. For example, if mangoes happen to be cheaper at the moment in Kuala Lumpur than in Sydney (adjusting for costs of transportation and the like), the arbitrageur will be able to profit from that 'spread' by buying low and selling high, lubricating out the market friction that had caused the price difference in the first place.

** Such as options (a contract allowing you to buy something at a stated price for a stated period) and futures (a contract entitling you to buy or sell something at a stated price at a stated future date), both of which in turn can be bought and sold at a market value that varies from time to time.

*** Secondary markets repackage, usually in bulk, and resell commodities or contracts originally sold in a primary market. For example, you may borrow money from a bank, but the bank then resells it and other loans as a package traded in a secondary market. Secondary mortgage market-makers include such well-known US institutions as Fannie Mae and Freddie Mac. An initiative by the US Department of Energy, in collaboration with leading financial houses, has recently made possible such a secondary market in energy saved by retrofitting buildings. Just like a mortgage, the savings can now be 'packaged' under standard rules, 'securitised' and sold on into a secondary market. New loans to finance the savings can be originated as quickly as the loans can be remarketed to secondary buyers. This makes it unnecessary for the buildings' owners to have the capital themselves: the retrofits can instead be financed entirely externally.

current market conditions – basically, the relationship between supply and demand. (Price is the signal that describes that relationship, and hence instructs market actors whether to consume or produce more or less of that commodity to help bring its supply and demand into balance.)

BIDS AND AUCTIONS

For example, in the 1980s, when Central Maine Power Company wanted to help its big industrial customers to save energy more cheaply than the company could make it, it invited them to bid for part of an annual fund of cash grants – to go to those companies that offered to save the most electricity per dollar of grant. The first year, one particular paper-mill manager put in proposals for electronic controllers for his big motors, and got all the money. The next year, his competitors bid too, and the programme was off and running. The utility screened bids in two stages: first, who was willing to save electricity most cheaply, and second, who among those winners was willing to pay the largest part of the cost themselves to stretch the grant dollars as far as possible.

Soon the 'industrial modernisation grants' grew into 'all-source bidding', later practised in some eight states.* If a utility wanted more electricity, it would run an auction, asking 'Who wants to make, save or displace electricity at what price? We'll take the low bids.' The low bids were generally savings. For the first time, ways to make and ways to save electricity could directly compete in the same market – not just in the same planning process, but head-to-head in a real market.

TRADING NEGAWATTS

If saved electricity can be bought and sold in this way, then it is 'fungible' like other commodities, able to be converted into money and back and to be traded across time and space. But then it can be traded among a variety of parties:

- Utility A can pay Utility B to save a certain amount of electricity and sell it back to A. (Contracts for this purpose have already been signed in the US.) This may become quite possible in Europe, since it is typically much cheaper to save electricity in, say, Eastern than in Western Europe, but equally

* However, more than 27 states have run supply-side-only auctions in which ways to produce electricity can compete against each other, but ways to save it cannot.

valuable in an interconnected grid. Moreover, in principle, though not yet in practice, such transactions, which free up transmission capacity (because saved electricity is already at the load centre and need not be delivered to it over the wires), should attract some sort of transmission credit, because they free up limited transmission capacity for valuable transactions by others.

- Saved electricity can also be traded for values other than money. (Thus the Pacific Northwest sells surplus hydropower to Southern California when it's most needed to reduce smog formation, and Southern California sells surplus nuclear power to the Pacific Northwest when it's most needed to float spawning salmon through the hydroelectric system. The amount of electricity both systems have in surplus depends in turn on how efficiently their customers use the available supply, so the trading arrangement could readily include efficiency trading as in the previous bullet.)
- Customer A, wanting cheap electricity, can save electricity by directly investing to improve Customer B's facilities or operations. The utility connecting the customers, acting as a 'negawatt broker', can then sell to A the same amount of electricity saved on B's premises, but at a discounted price, so that part of the saved operating cost rewards A while the rest goes to all customers and to the utility's shareholders.

All the methods described here, both for marketing negawatts and for making markets in negawatts, can apply to a wide range of resources, not only to electricity. So let's pick an example of 'wheeling' savings between customers, but this time for water.* Morro Bay, California, was short of water in the late 1980s. The town authorities therefore told the homebuilders, 'If you want to build a new house here, then you must first save, somewhere else in town, twice as much water as your new house is going to use.' The homebuilders, wanting permits, went door-to-door installing water-saving appliances in one-third of the town's entire housing stock in the first 2 years. Now imagine how this could work: Someone comes to your door and says, 'Just look at this wonderful new toilet. It's so beautiful – designed by a Swedish sculptor – that there's one at the Museum of Modern Art in New York. It uses

---

* Scores of water examples are given in Rocky Mountain Institute's *Water Efficiency: A Resource for Utility Managers, Community Planners and Other Decisionmakers*, Publication no. W91-27, fourth printing, 1994, prepared under a USEPA Cooperative Agreement.

only one-seventh as much water as your present toilet, but it's more reliable, quieter and works better. And I'd like to give it you and install it for free.' But being a shrewd businessperson, you reply, 'Yes, I heard you chaps are doing that. How much is it worth to you? How much do you want to pay me to let you install that thing in my house? How much do you want your building permit?' Then you discover what saved water is worth, because you have just made a market in it. And you can do the same thing for saved electricity, wood, gas, cobalt or any other commodity.

Savings can even be traded across national boundaries. In the 1980s, the Canadian utility Hydro-Québec wanted to build a gigantic hydroelectric project, La Grande Baleine, at high economic, environmental and cultural cost. Of its output, 450 MW was to be sold to Vermont utilities at a price which, over time, worked out to about 9 cents per kilowatt-hour. That proposed price, based on implausible projections of ever-rising oil prices, was far too high to interest Vermont. But Richard Cowart, head of the Vermont equivalent of the PUC, had a better idea to offer Hydro-Québec: 'Let us in Vermont come to Montréal and help you improve your buildings and factories to save 450 MW, which we'll then buy back from you for, say, 3 cents per kilowatt-hour. It'll cost us less than 1 cent to save each kilowatt-hour and three to buy it back; that's less than 4, which is much less than the 9 cents you wanted, so we like that idea. But you, Hydro-Québec, will make more money this way, because the power we'll save and buy back is from an old dam you paid for 20 years ago: it's pure profit. Rather than your building a costly, risky dam up north, we'll help you build a very cheap, no-risk dam in Montréal. It'll make your provincial economy more competitive. In short, Hydro-Québec, your power is worth far more saved and resold than wasted at home.' The proposal was rejected, because Premier Robert Bourassa was actually interested in dams, not dollars; but it was still a good idea, and the market rationality it reflected finally came home to roost when a new Québec government in late 1994, chastened by evaporating markets for La Grande Baleine's costly power, shelved the project.

Note that, in the Québec example, the 'spread' available for arbitrage is essentially the difference between the cost of making the Grande Baleine electricity (9 cents including profit but excluding environmental and social costs) and the cost of saving the same kilowatt-hour (less than 1 cent). If, therefore, 3 cents isn't the right price, there's a lot of room to negotiate. Indeed, the business opportunity here – capturing the spread between the cost of negawatts and megawatts – must be one of the most lucrative on the planet.

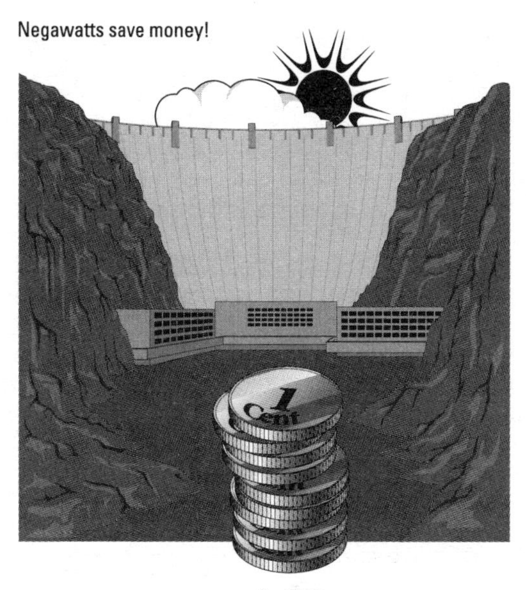

Figure 18 Negawatts can save a lot of money. The megawatts from the planned gigantic dam of La Grande Baleine in Québec would have been some nine times more expensive (9 cents per kilowatt-hour) than the negawatts with whch they had to compete (1 cent per kilowatt-hour). In the end the dam was not built.

For example, in contemporary Europe, it does not require much imagination to see that cheap Norwegian hydroelectricity, now used very wastefully, could be cheaply saved, then sold on at a higher price to replace, say, retiring nuclear plants in Sweden and Germany or dirty coal plants in Germany.

To help facilitate such transactions, you might expect to see the emergence of spot, futures and options markets in saved electricity. This is no longer far-fetched: an electricity futures market is already in limited operation in Britain, and both spot and futures markets in avoided power-station sulphur emissions have already been made on the Chicago commodity exchange. A 'negawatt future' would be an underwritten contract, saleable for a market value, to deliver a certain amount of saved electricity at a specified time, place and price. It can be measured; therefore it can be sold.

It is also possible to buy back from the customers either avoided electricity demand or reduced uncertainty of electricity demand. To a utility portfolio planner or a power broker, these are valuable commodities that deserve markets in which to express their risk-hedging value. For example, one could attach to land or a building a restrictive covenant* promising never to use more than *x* megawatts. That is useful, and could be sold to a utility that then needn't go on planning as if the property were always going to use more and more, and hence can avoid investing in capacity to serve that assumed growing demand. This is somewhat akin to the US Environmental Protection Agency's 'bubble concept': someone wanting to open a polluting factory in a place with dirty air must first abate that much pollution somewhere else in the same airshed. In practice, this is done not by physically finding and cleaning up (or shutting down) another plant, but rather by buying from a broker, at market value, a previously earned certificate of decreased air pollution. Buying stabilised or reduced electrical demand should hardly be very different. In fact, a special case of this is already widely practiced: the 'curtail-to-threshold' tariff, under which a customer expresses willingness, in rare power emergencies, to receive a reduced supply of electricity down to a certain 'threshold' mutually chosen in advance, and receives a monthly payment from the utility for that willingness, whether the curtailment actually occurs or not.

CROSS-MARKETING

Another intriguing feature of saved resources is that many diverse parties can offer them for sale in various places. For example, electric utilities normally have a 'franchise monopoly' entitling them, but nobody else, to sell electricity within a certain service territory. But there is no franchise monopoly on negawatts: they may be freely sold anywhere. Utilities can therefore sell negawatts** in each others' territories, and about a dozen have already done so. For much of the 1980s, Puget Power Company sold electricity only in its Washington State territory, but its unregulated subsidiary sold electrical savings in nine other states – wherever tariffs were especially high. In the eastern part of Germany, where a bizarre law promulgated by the three big western generating companies restricts the

* Or what is called an 'easement in gross' where state law so permits.
** In practice, this could mean selling any required mix of advice, design services, project management, financing (usually the most important part), equipment, installation, commissioning, operation, maintenance and monitoring. The key elements are usually information, coordination and financing.

municipal distribution companies from competing in generation, they are not restricted from offering efficiency services, and could easily undercut the generating monopolies that way: similarly in Japan or anywhere else where generating monopolies still suppress competition.

Even more disquieting to some electric utilities is the concept that *gas* utilities can sell *electric* efficiency: they are just as good at delivering it, typically to the same customers, but have no inhibitions about reducing the sale of electricity. In fact, doing so could also change the operation of buildings in ways that help companies to sell more gas. Consider, for example, the common North American situation of a big office building that is lit and cooled by electricity but heated by gas. The gas company could offer to provide the engineering, financing and project management to retrofit the lighting system: it could guarantee the electrical savings, especially reduce the peak cooling loads on hot summer afternoons, help people see better, make the building more attractive and incidentally sell slightly more gas because the more efficient lights will provide less fortuitous heating in the winter. But then the gas company can propose to remove the electric chillers altogether, substituting (and leasing or operating if desired) a gas-fired or solar model or even a gas-fired cogeneration apparatus that provides electricity, heating, hot water and cooling. The alternative cooling and/or generating equipment won't be big and costly, because both the lighting and the cooling loads have been greatly reduced by the lighting retrofit, and a gas-fired chiller won't use much gas, because now the cooling load is much smaller and the cooling season much shorter. Thus the electrical saving has shifted the balance of advantage towards gas.

Of course, there are examples in both directions, and for all kinds of customers. Smart regulators, seeing in this a potential to foment competition that cuts costs for all customers, will wish to reward electric and gas utilities for their success in the negawatt *or* negajoule market without discrimination – rewarding either for cutting customers' bills of either kind, so that both will vigorously compete in their own and each others' markets. In all, these and other ways of both marketing negawatts and making markets in negawatts amount to an extremely powerful menu of options that can be selected and adapted to fit virtually any circumstances. They unleash bounty-hunters to go find inefficiencies anywhere and exploit them.

And they have lessons far beyond electric and even water efficiency. Could not the social value of an unburnt barrel of oil, for example, be reflected in prices signalled by spot, futures and options markets in saved oil, so that arbitrageurs can exploit the spread between the cost of barrels and negabarrels? Could not 'negatonnes' of saved or recycled metals be traded alongside tonnes of virgin metals? Might not one firm or country pay brokers to find others with whom resource efficiency improvements and pollution abatements could be most profitably traded? How much is it worth paying people to stay off the roads so we needn't build and mend them so much and suffer their pollution and delays? As with subatomic particles, for every resource there is an equal and opposite 'antiresource', for every activity an abatement, each arguably meriting a value and a market.

Making markets in resource efficiency and environmental protection has even wider implications. People don't want barrels of oil or raw kilowatt-hours; rather, they want the end-use services that the energy provides. But if the only way they know to get the desired services is by buying electricity, then a choice to use less electricity and more efficiency cannot be expressed in the market; utilities will therefore have an effective monopoly in providing the final services; and demand for electric efficiency will appear to be very small. The solution is to articulate, make available and make markets in the efficient use of electricity, so that electricity must openly compete with electrical productivity.

Analogously, people don't want weapons; they want to be safe and feel safe. But if the only way they know to get security is through weapons, then the weapons vendors will have an effective monopoly in providing security services, and demand for weapons will appear to be very insensitive to their price (reinforced by the monopoly, monopsony and often corruption of the political process through which the weapons are bought). The solution would be, as Rocky Mountain Institute is attempting, to articulate and bring to market the specific, practical elements of an alternative concept of security – so that other ways to achieve freedom from fear of privation or attack will become more available as alternative intermediate goods with which weapons must compete in the political marketplace.

Properly pricing depletion and pollution cannot replace a proper regard for our moral obligations to beings in other places, times and forms. But if imaginatively combined with flexible and accessible markets, prices can at least apply to corrective mechanisms the same vigour and ingenuity, the same genius of the market, the same diversity and adaptability that have got us into this mess. Markets that apply the new price signals, and focus them into action, will at least help to make the struggle between destroying and creating the future into a more nearly equal contest whose issue is less in doubt.

Chapter 6
Reward What We Want, Not the Opposite

6.1 Correcting Perverse Incentives

'Least-Cost Planning', in Chapter 5, described the importance of forming electricity prices (and similarly for gas, water and so on) so that regulated utilities are rewarded for reducing customers' *bills*, not for selling them more of the commodity. This reform reflects a belated realisation that all regulation is incentive regulation; the only question is which kind of behaviour or result is being rewarded or penalised. Utilities, like other firms and individuals, are remarkably good at following the incentives they're given – however irrational. The relatively slow progress of saving electricity in most countries is clearly related to the continuing practice in some 46 of the United States, as virtually everywhere else in the world, of increasing utilities' profits when they sell more electricity and decreasing their profits when they sell less. Naturally most of them are not too enthusiastic about selling less. Some even have efficiency programmes (to help themselves look good to the regulators) but then more than undo them with even bigger efforts to increase electricity sales again!

But what about the vast majority of ordinary industries, whose prices are not set by a regulatory commission? Most industries in the market economies set their own prices, and few would wish otherwise. Does the free-market formation of their prices mean that they have a rational incentive to sell their goods and services more cheaply to gain market share and boost profit margins?

Not exactly. In fact, all too often, not at all – as Rocky Mountain Institute began to realise when preparing, for its E source subsidiary, an encyclopaedic *Space Cooling Technology Atlas* in 1992. The fifth book of a six-volume set, this atlas described how to save about 80–90 per cent of the energy now required for space cooling and air handling in US buildings, yet provide better comfort. (Even better results have since been shown in practice, as noted in Chapter 1, pp 15 and 58) But towards the end of the writing, a curious question occurred to Amory Lovins: how much capital has the US already misallocated by not designing buildings in the optimal fashion described in the atlas? That suboptimal design, he found, has caused the needless installation of some 200 million 'tons' of cooling equipment and some 200,000 peak megawatts of peak power supplies to run it – two-fifths of the peak load for all purposes in the whole country! At replacement cost, that wasted equipment would cost close to $1 trillion ($1 million million).

WASTING A TRILLION DOLLARS JUST ON US AIR-CONDITIONING INVESTMENTS

How on earth could a mature market economy misallocate a trillion dollars just on air-conditioning? Surely this bespeaks a more radical market failure than even the most jaded market critic might have imagined. But on close examination, it turned out (Lovins, 1992) that there was a stunningly simple explanation. Each of about two dozen parties involved in the building process had perfectly perverse incentives: everyone involved was systematically rewarded for inefficiency and penalised for efficiency. If we had set out to design a system of incentives and institutional structures to make buildings use about ten times as much energy as they should do, be less healthful and comfortable than they should be, and cost more to build than they should do, it would be hard to improve on the system we've actually got – in the US, the UK, and virtually worldwide.

This applies to all the identified people and organisations involved in conceiving, approving, financing, designing, building, commissioning, operating, maintaining, selling, leasing, occupying and renovating buildings, both commercial and residential. Let's consider a simple example: the incentives seen by the architects and engineers who design, say, a big office building.

Whether by traditional practice, as in the US, or by formal rules, as in Germany, design professionals in virtually all countries are paid a fee based on or negotiated from the *cost* of the building or equipment they specify. Imagine that you are the engineer designing the cooling and ventilating equipment and other kinds of building services for a big new office building. The normal way of 'designing' that equipment is not well optimised; rather, it's based on slightly modifying previous drawings for things you know will work (so you won't get sued), generously sized to deal with whatever lights, windows, office equipment and other cooling loads the other designers are meanwhile independently specifying. Lee Eng Lock (see Chapter 1, p 53 on his efficiency achievements) irreverently, but not so inaccurately, describes this Mechanical Engineering Standard Operating Procedure, this 'MESOP Fable', as:

- ❑ Take previous set of drawings.
- ❑ Change project box (which says which project it is).
- ❑ Submit drawings to client.
- ❑ Building is constructed.
- ❑ Client complains of discomfort.
- ❑ Wait for client to stop complaining.
- ❑ Repeat process for 25 years.

This safe but sloppy procedure, which a Canadian engineer has called 'infectious repetitis', makes your mechanical equipment so big, complex and costly that the traditional design-fee percentage of its cost is about enough to make a living on, especially since you're not working terribly hard to make the design refined or innovative. It will work, it will consume a lot of capital and energy, and you won't pay those costs.

But suppose instead that you put the dis-integrated design team back together, and treat the design process as a team play, not a relay race. Instead of having each designer throw the drawings over the transom to the next – the way the architect normally throws you the drawings for a pretty 'all glass and no windows' box and says, 'Here, cool this thing somehow' – all the architects, engineers and for that matter landscapers, artists, constructors, occupants and maintenance staff work together in a joint design process, as they did in the exemplary buildings described in Chapter 1, pp 10–29 and 58.

Now the superwindows let light in but block unwanted heat, the form of the building puts the heat only where you want it, the minimal artificial lights and the office equipment become extremely efficient and the whole building needs scarcely any mechanical equipment, rather like the Queen's Building in Leicester (Chapter 1, p 24). What happens now? The building's owners and occupants benefit from lower construction cost, lower operating cost, and greater amenity and productivity – but you go bust. Why? Because the mechanical equipment on whose cost your fee was based is now much smaller, simpler and cheaper, and a lot of it disappeared altogether. Even if your fee were already fixed in advance, you're working harder to do real engineering optimisation, but for the same fee. Either way, your profits become negative. You just got penalised for doing excellent, resource-saving, cost-reducing engineering.

GIVING DESIGNERS A PIECE OF THE ACTION

Fortunately, there are solutions to this problem, as to all the other perverse incentives in buildings. For example, design professionals could be rewarded for what they save, not what they spend: they could earn their traditional fee for 'plain vanilla' design, but in addition be given a certain percentage of whatever reductions in life-cycle cost they achieve compared with some accepted baseline design. (Both the tools for calculating energy and capital costs and, in most countries, the baselines are well established and fairly easy to use.) Rocky Mountain Institute is currently conducting an experiment to test this approach empirically in several major building projects in the US. Architectural associations in the US, Germany and Switzerland are starting to work out rules for this 'shared-savings' approach, and some individual practitioners are already starting to apply it to specific projects. Obviously, it is in the interest of smart designers to educate their clients to demand, and for their own firms to bid, a two-part fee (standard basis plus *x* per cent of savings achieved by extra effort). Whichever design firms and clients succeed early in capturing the benefits of this approach will get their competitors' attention, enabling competitive forces to spread the idea automatically.

Another possible solution is for the electric utility to reward the designers directly with a rebate equivalent to, say, 3 years' worth of energy savings. (Ontario Hydro, the giant Canadian utility, actually started to do this a few years ago, though the programme and most of its other efficiency efforts then got suspended during a wider fiscal crisis.) The rebate could be paid half up front, when the design costs are incurred, and half a few years later, trued up to the actually measured savings, so that the designers have a strong incentive to ensure by careful supervision that their intentions are fully carried out in construction, commissioning, training and operation.

Why would utilities want to pay for better design? Because the savings are the cheapest resource for their system to acquire – they represent electrical services delivered at much lower cost and risk than by generating more electricity, even in existing power stations. Today, most US utilities offer cash rebates to customers who buy more efficient equipment (better chillers, fans, pumps and so on). But if, as RMI's and E source's analyses clearly show, superefficient equipment (when properly sized and its benefits properly counted) may actually cost *less* than normal equipment – but isn't being widely used because few designers know how to choose and combine it and because they'll be penalised if they do – then utilities can get more 'bang per buck' by rewarding good design than by rewarding the purchase of efficient equipment. In fact, by paying relatively small rebates to designers, they may well be able to eliminate their costlier rebates for equipment, because now the designers will be rewarded, not punished, for choosing cheaper and more efficient equipment.

These approaches offer designers an opportunity to double or triple their normal fee – a good way to get their undivided attention, to distinguish firms good at integrated resource-efficient design and to reward fairly the hard work that such design requires. Conversely, honest and careful estimates can be encouraged by procedures akin to one used in retrofitting buildings for the Singapore government: if savings fall short of what was promised, the responsible parties must write the government a cheque equal to 10 years' worth of the shortfall. One proposed project in California intends to combine these two approaches: the designers will get to keep x years' worth of energy savings (in excess of a level 10 per cent better than the already strict state energy standards), but if they don't achieve what they promised, they'll pay y years' worth of the shortfall as a penalty. This combination can fairly combine risk with reward, making all the parties better off.

Cleaning the Augean Stables

If we correct the perverse incentives now seen by mechanical and electrical engineers and by the architects for whom they consult, we'll have solved one problem for three parties. They'll still have other problems too, each requiring specific actions to correct (Lovins, 1992), ranging from inadequate design education to peculiar rules of professional practice to inadequate coordination with constructors and building operators. But even if we solve all those problems, we'll still have more than 20 other parties to the building process who still have their own perverse incentives. Any of those parties can be a show-stopper. Thus, making markets work properly to produce cost-effective buildings is an unusually complex institutional problem requiring sustained and concentrated attention by practitioners, their professional societies, their regulators, other public-policy bodies and other market actors.

For example, most commercial and many residential buildings are leased, not owned by their occupants, so the lease arrangements must be rewritten so that both owner and tenants share in the savings either achieves and both have an incentive to achieve them. Leasing brokers need to calculate buildings' pro forma economics based on their actual performance, not rule-of-thumb assumptions, so that efficient buildings get the market credit they deserve. The way builders choose, buy and install equipment needs comprehensive reform. Buildings need to be independently 'commissioned' so that they actually work as intended, and then their operators (and often occupants) need to be trained to keep them working that way. Building operators need more resources and self-esteem. The various parties involved in financing buildings need to be rewarded for saving money, not for making deals or acting quickly. From beginning to end, the building process begs for fundamental and systematic reforms. Without them, the US continues not only to add to its trillion-dollar misallocation to air-conditioning, but also to expand its $12-trillion stock of buildings that were obsolete before the doors opened. And so does every other country in the world continue to add to its own misallocations.

The benefits reaped for rewarding the outcomes we want (buildings that are cheaper to build and run, work better and look nicer) rather than the ones we don't want (*e.g.* costlier buildings and equipment) are immense. Often they come back most of all to ensuring good design in the first place. In a typical US office building, over the years we pay the workers about a thousand times as much as we paid once to the mechanical engineer whose handiwork, the comfort systems, largely determines how productively the workers can operate and whether they stay happy enough to keep paying the rent. If better engineering education improved the efficiency of building services by only 20–50 per cent (as compared with the 70–90+ per cent illustrated in Chapter 1), just the avoided utility investment, present-valued over the career of a single mechanical engineer, would amount to some $6–15 million per brain – surely a hundred or a thousand times more than the better engineering education would cost. And since that engineer's lifetime design portfolio will serve on the order of 64,000 office workers (say) with present-valued salaries of $35 thousand million, the 6–16 per cent gain in labour productivity observed in well-designed 'green' buildings (Chapter 1, p 29) makes the benefit of the better engineering education on the order of 1 million times bigger than its cost. Where else in society can we gain such enormous leverage of vast benefits from small costs?

6.2 Responsibility Requires Responsiveness to Feedback

Often the task is far simpler than correcting perverse incentives; it's simply adding basic incentives to do something sensible. For example, many factories sited on riverbanks traditionally drew their water supplies into one pipe and discharged their wastes out another. Where this arrangement still exists, why not simply put the intake downstream of the outfall, so that whatever pollution the plant emits, it gets right back again (see Plate 12)? In an extreme case, why not simply connect the two pipes? If wastes are OK to put into the river whose water others use, why aren't they OK in the water that the plant itself uses? Some 'ecoteurs' have exploited this logic by 'recycling' a polluting factory's discharges into its own premises – a sure way to get the management's attention.

If a laboratory building is discharging hazardous solvent fumes from its fume hoods up a stack so that they are diluted and rebreathed by everyone else, why not press the operators to explain why they are willing to make everyone outside breathe what they are unwilling to breathe themselves? This logic rapidly leads to zero discharges, usually through changing the process to eliminate the solvents or, failing that, to full recovery of the valuable solvents – often costing less than replacing them and dealing with their 'safe' discharge in the first place! This probably isn't a universal prescription, but it's far simpler than thick volumes of impenetrably intricate regulations, and often leads to a better and more profitable design solution.

If people are concerned that an oil company is not safely operating a refinery or a petrochemical plant, why not ask it, as a token of its good faith, to require that all the plant's senior managers, and their families, live at the downwind site boundary, so that they are the first to be exposed to whatever releases occur?

If a maker of nuclear power stations claims its new plant designs will be perfectly and 'inherently' safe, why not ask it to waive its legal exemption from unlimited liability* for accidents caused by that plant? After all, if it's really safe, the company has nothing to worry about. Or are they unwilling to bear themselves the supposedly nonexistent risks to which they wish to expose the public?

The logic of such simple examples is clear enough – and should be more frequently applied.

6.3 MAKING BETTER CHOICES POSSIBLE IN THE TRANSPORT SECTOR

Another simple way to get better results is to introduce competition between alternative solutions. Consider, for example, the simple matter of employee parking. A third of all US household road mileage is for commuting to work, where an estimated 96 per cent of driving employees park free in spaces requiring up to several times the square footage of their office space: the free employee parking lot is a universally expected perquisite, and tax-free to boot. But suppose that the employer, either voluntarily or by law (local, state or national), instead charges fair market value for the parking spot – following sound market principles – and pays every employee a 'commuting allowance' of equal after-tax value.

Now an employee can continue to drive to work, pay for the parking space and incur no net gain or loss: it's the same as before. But it will occur to many workers that if they get to work in any cheaper way (ride sharing, jitney, public transport, walking, bicycling, living nearby) or don't need to (part-time telecommuting, out-of-body trips and so on), they can pocket the difference. By injecting money into the system, the employer has fostered and monetised competition between all modes of access to work. Now that they can start to compete, it's easier for the best buys to win. But if everyone simply gets a free parking space that can't be 'cashed out' – that isn't fungible for other options – then no competition can occur, and cars will continue to dominate commuting.

---

* Most countries, following the example of the US 'Price-Anderson Act', have passed such laws, which provide that above some nominal level, nobody is liable for nuclear accidents, even if caused by gross negligence, and the victims (who, after all, cannot obtain private insurance against such risks) must bear their losses. This can hardly be considered a good market incentive for safe operation. In fact, Westinghouse, seeking Export-Import Bank subsidies to work on the Temelín reactor in the Czech Republic (as a precedent for other Eastern European work), is seeking an extension of American taxpayers' liability to include accidents in the Czech Republic, even if caused by other participants in the project. As might be expected, this desire seems to present legal and political difficulties.

Where will the employer get the extra money to pay enough commuting allowance to cover both the parking space and the worker's income tax on it ($15–20 thousand million a year to the US Treasury)? Simple: when parking has to compete with all other modes of access to work, not as many people will want to drive and park so many individual cars. If the marginal tax rate is (for illustration) 25 per cent and the competition decreases the demand for parking by 20 per cent, the employer will break even. If demand for parking declines even more, as some early evidence suggests it will, then the employer actually makes a profit by being able to sell or lease its now-excess parking spaces (at fair market value), or by not having to build so many of them. The wealth created by the more efficient access solution is then split between the employer, who pays less to provide parking, and the workers, who get cheaper access to work.

California's South Coast Air Quality Management District already requires this sort of 'cashing-out' option for employee parking, and President Clinton's 1994 global-warming strategy moves in the same direction. So does the 'information superhighway' in an era when half of Americans work in the information economy. Indeed, information itself (not counting services) is a bigger business than the manufacturing, energy and materials industries combined. Even such an extreme measure as bringing optical fibre into every home would cost less than we spend on road building every couple of years.

The parking-cash-out idea can be extended in many ways. In much of the US, for example, zoning authorities require that every new flat or condominium unit come with a parking place. (That's the opposite of the system in, say, Frankfurt, where you can't build an office with associated parking: workers must buy their own; or in Tokyo, where you're not even allowed to buy a car unless you can prove you have a place to park it – no small feat at local land prices.) Now suppose that real-estate developers were forbidden to provide a parking place with each housing unit, and instead were required to supply a perpetual transit pass with each unit, as was proposed in San Jose, California. Developers like this idea, because an annuity to buy a perpetual transit pass costs many times less than a parking space (which costs around $15,000–35,000 in most American cities), and they can keep the difference.

This is further akin to a Stockholm proposal that downtown residents who wanted to drive during a given month be required to buy a special permit – which would also be their free pass to the regional transit system for that month. Having bought it, they might as well use it, and even if they didn't, their payment would help provide better transit service for all, helping public transport compete with cars, most of whose costs are socialised through taxes rather than directly paid by drivers.

The American automobility problem is rooted in poor land use, mandated by obsolete zoning rules left over from an era of dirty and dangerous smokestack industries. To guard us from living near these rapidly vanishing public nuisances, current zoning mandates land-use patterns that maximise distance and dispersion, forbid proximity and density, and require universal car traffic on wide, highly engineered roads. In the name of amenity, it makes every place costly, polluted and unliveable. To paraphrase Swiss electric-car pioneer Uwe Zahn: Cars need roads, displacing pedestrians, who then use more cars and demand more roads, on which more cars operate, impairing public transit and making it unattractive, so transit passengers also switch to cars, which make the city noisy and polluted, until city-dwellers migrate to the countryside where they too need cars, for which still more roads are built.... In contrast, modern land use discards segregation by income and use, reintegrates activities, creates pedestrian-based neighbourhoods and preserves the ability to drive only where it's really needed. This creates compact, pleasant, popular and profitable developments with less crime, sharply higher property values and cohesive communities. Housing is often near or next to light commercial uses and the clean, service-based industries that now dominate the industrial world's economies.

All these benefits can be obtained by educating developers and zoning officials, but they can also be accelerated by, for example, using existing systems of density bonuses and penalties to reward the colocation of housing and workplaces, or to reward proximity to public transport. Since the 1950s, such simple techniques have helped steer nearly all of Toronto's development to within a 5-minute walk of Underground or light rail. Recent California comparisons suggest that, over little more than a decade, such incentives for clustering can so shift land-use patterns that a person-mile of transit can displace 4–8 person-miles of car travel. Arlington, Virginia, already slashes traffic by using Metro (tube) stations as development foci. Indeed, whenever a new Washington-area Metro station opens, real estate values jump 10 per cent for blocks around, focusing further private development – $650 million worth just in the system's first 3 years.

A long-standing subsidy to sprawl, mortgage and tax rules that encourage highly dispersed suburbs, could be corrected by 'commuting-efficient mortgages'. In the US, where the secondary mortgage market is dominated by two federally chartered corporations, Fannie Mae and Freddie Mac, their rules have been changed in recent years to qualify energy-efficient homes for a bigger mortgage on less income, because their low energy costs can support more debt service with less risk of default. This sensible approach is well established. But now David Goldstein, senior scientist at the Natural Resources Defence Council, has suggested that including a neighbourhood's typical commuting costs in the same formula would make urban housing cheaper and suburban sprawl costlier, better reflecting their social costs. How readily typical workers in a given neighbourhood could reach their jobs – depending on their proximity to jobs and to public transport – can easily raise or lower their disposable income by hundreds of dollars per month, and if used for debt service and thereby leveraged tenfold, that could enormously increase the number of families that could afford to buy their homes. In 1995, Fannie Mae launched a $1-thousand-million pilot programme to test this approach. If successful, it could provide a strong incentive for urban infill and against suburban sprawl.

Even the way we buy motor fuel can be made more pregnant with price information, especially in the US, which insists on keeping prices far below full costs California resource policy consultant Mohamed El-Gasseir has devised, and famous American financial advisor and broadcaster Andrew Tobias is promoting, an innovative way to help signal American petrol's social cost while reducing everyone's bills. It's called 'pay-at-the-pump' car insurance, and its principle is simple. Most Americans now pay more per mile for car insurance than for petrol. Most of the insurance is related to collisions, whose risk increases with miles driven. States can therefore unbundle the insurance premium and charge the collision-related part at the petrol pump, then forward it to the private insurance companies in proportion to their market share. The remaining insurance premium, for theft and casualty risks, would be paid through the post to your chosen insurance company in the usual way. A truing-up term on your bill would reflect coverage, competitive and actuarial differences. (Insurance could also be made no-fault, paying the injured rather than the lawyers.)

Under pay-at-the-pump, the apparent price of petrol would rise by perhaps 30–80 cents per US gallon (about 5–14 pence per litre) – still about the lowest in the industrial world, but a more accurate price signal than now. Yet the increase is not a petrol tax; on the contrary, people's total cost of driving would go *down*, because there would be no socialised accident costs from uninsured motorists (now perhaps a quarter to a third of all American drivers): everyone who buys motor fuel automatically buys collision and liability insurance at the same time. This is simply a smarter way to buy insurance – and reminds us, whenever we fill up, that insurance is part of the cost of driving. And it's not even new. Hungary and, in an indirect way, New Zealand have used a similar system, though both have lately abandoned it, apparently in the interest of harmonisation with less advanced neighbours. It seems they were simply too far ahead of their time.

These and many other innovations are the building blocks of an emerging end-use/least-cost policy framework and decision process to foster fair competition between all modes of access – including those that displace the need for physical mobility, such as already being where we want to be, and thereby not solving the transportation problem but avoiding it. Creative public-policy instruments can introduce market mechanisms to a transportation system long crippled by lopsided subsidies and top-down central planning by an almost Stalinist approach in which the road planners say, 'You will have highways here and parking there and car-based infrastructure all over, and you will pay for it through your taxes. If you want to do something else – well, of course, it's a free country; you can get round however else you wish, and pay extra for it.' That is not true competition nor a level playing field. Its social costs are proving ruinous even for the wealthiest nations, whose unsupportable urban congestion is largely caused by the overprovision of apparently free urban roads and parking. But needed innovations are starting to emerge: ways to make parking and driving bear their true costs, improve competing modes and substitute sensible land use for physical mobility.

The multitude of creative ways to help markets work better in mobility, access and transportation invites irreverent speculations. Why not simply make markets in 'negamiles' and 'negatrips', so we can discover what it's worth to pay people to reduce their driving? If anyone could make money from any way to get access that's socially cheaper than driving cars, wouldn't we all drive a lot less?

For that matter (asks Douglas Foy, who directs the Conservation Law Foundation of New England), why not privatise each transport mode into one or more regulated public utilities that are rewarded, like modern electric utilities, not for providing a bigger volume of service but for minimising users' total social cost? Now that automatic electronic billing can charge drivers their true social costs as they whiz by (the tariff depending on the time of day, degree of congestion and so on), we could eliminate all transport-related subsidies, make each mode of transport pay its own way and convert all modes of travel from a tax burden into a stream of sale payments or royalties back to the public sector that built the infrastructure. Perhaps this is too radical an idea for a society built on corporate socialism for cars and free enterprise for other modes. But perhaps, too, it might work a lot better than present arrangements, where America increasingly looks like the place where foreign cars fueled by foreign oil drive over crumbling bridges and potholed roads to places scarcely worth going to.

6.4 Making Prices Tell the Truth

The problems of excessive automobility, like many others in our society, are caused not only by lack of real competition between alternatives but also by untruthful prices. We need economic *perestroika* built on economic *glasnost*, for if our prices tell lies, they cannot guide true choices, and if choices are not available, prices hardly matter anyway.

In America, for example, the social costs of driving – related both to the conversion of fuel into smog and to congestion, lost time, accidents, roadway damage, land use and other side-effects of driving itself – are largely socialised. 'External' (or, as Garrett Hardin called them, 'larcenous') costs approaching 1 trillion dollars a year, perhaps a seventh of the American GDP, are borne by everyone but not reflected in drivers' direct costs. In this Alice-in-Wonderland world, no wonder the benefits of cleaner and more efficient cars have lately been swallowed up by even more driving in ever more cars.

Yet a few existing proofs show that proper pricing works. Singapore, whose prosperity could have made it another bumper-to-bumper Bangkok, is rarely congested, because it taxes cars heavily, auctions the right to buy them, imposes a $3–6 daily user fee on anyone driving in the city centre and puts the proceeds into excellent public transport. This level playing field, where cars and transit both pay their way, yields a liveable city. People who want to drive can, but they must pay their own costs, improving both their own choices and others': they get what they pay for, and pay for what they get.

What Singapore does with police can also be done automatically. The same electronic gadgets that let drivers in Dallas and Oklahoma zoom through toll-gates, debiting their accounts automatically (and at one-quarter the cost) without violating privacy by disclosing their location, can be used to charge more for parking – and for roads, tunnels and bridges – when they're most congested, just as we already charge for phone lines and electricity depending on their scarcity at the time. Oslo and Bergen do this with traffic right now. A similar system for thousands of fleet vehicles at Los Angeles International Airport, automatically surcharging too-frequent visits, has cut kerbside congestion 20 per cent and more than doubled revenues despite lower fees. Using the same technology that bills you depending on when and where you're driving, drivers could even be charged the full social cost they now impose on others – perhaps quadrupling drivers' cost of urban car commuting and thereby virtually eliminating even the most intractable urban congestion.

Of course, the list of price distortions needing correction is longer than this book. American railway pricing that charges more to transport a tonne of recycled copper than of virgin copper; other forms of discrimination against resource efficiency and recovery; over $30 thousand million a year in direct federal subsidies to the US energy system, $100 thousand million to EU agriculture, and probably more worldwide to transport – all are the tips of a whole seaful of icebergs on which even the most prosperous economy can quickly founder. Estimates by business reformer Paul Hawken indicate that the direct waste in the US economy – from subsidies to do silly and uneconomic things, to remedial costs for damage done elsewhere, to uncounted depletion and pollution, to other follies too numerous to mention – constitutes at least half the GDP.

And the problem is not only false prices but also false price *structures*, as when a utility wishing to sell more energy provides volume discounts ('declining block tariffs') – the more you buy, the less you pay per unit – or, conversely, has a high fixed charge per customer, making effective per-unit costs astronomically high for frugal users and correspondingly lower for extravagant users. It is neither impossible nor even unusual to price a commodity at a reasonably accurate level on average, but to structure the price in a wholly irrational and counterproductive way.

If economic democracy and customer sovereignty are won by voting with our wallets, then we are living under a tyrannical cloud of demagoguery, fraud and electoral irregularities. Even the most conscientious votes cast with our money will not elect the results we thought we had chosen unless they go to the candidates described, bound to the platforms they advertised, and will be accurately counted and announced. If free and honest voting is the hallmark of a mature democracy, then free and honest pricing and buying are the hallmarks of a mature economy. It is not only in the painfully difficult reforms of Eastern Europe that we learn how impossible it is to achieve the one without the other.

6.5 Feebates

Suppose that many people are doing something silly that costs us all a lot of money, but that there's a simple alternative that would instead save us all a lot of money. What can we do to reduce the waste?

One way is to wring our hands, complain about how stupid those other people are and pat ourselves on the back for our own superior behaviour. This may make us feel good, but it doesn't do much to change others' behaviour, and anyway, self-righteous people aren't much fun to be around.

Another way is to pass a law or create a regulation requiring people to do something different. This is the classic approach of, say, building standards. If people are law-abiding, or expect strict enforcement (which someone actually has to carry out and pay for), then some reasonable degree of compliance can be expected, perhaps after some delay as people learn how to change their practices. On the other hand, they are unlikely to do better than the standard, especially if they think compliance is costly. That's why many builders proudly proclaim that their buildings 'meet code' – meaning that if they were any worse, the builder would be gaoled.

A further problem with standards is that they're static, but technology is dynamic. Standards don't improve by themselves as technologies do. In fact, because it takes some time to pass them, and usually there's a lot of negotiation to make compromises with groups that think they'll be hurt, most standards are obsolete before the ink is dry. That's why Britain is only now considering a rule that new houses have to be insulated better than piggeries, or as well as Scandinavia required about 30 years ago. And it's why Germany's new thermal insulation standard (*Wärmeschutzverordnung*) falls so far short of what's practical and worth doing (Chapter 1, p 13).

However, there's an exciting alternative to prescriptive regulations that are outdated, static, sometimes ignored and always inadequate. It combines the idea of rewarding the results we want, not the opposite, with the idea of sending more accurate price signals and letting people figure out their own ways of responding. It's a combination of a fee charged for inefficiency and a rebate rewarding efficiency – a combination, invented by Richard Garwin, that American energy-efficiency expert Arthur Rosenfeld calls a 'feebate'.

For example, suppose that new buildings that waste energy or water impose large costs on everyone else (as they do, because new power lines and generating stations, or new water supplies and wastewater treatment plants, are very expensive). Then a 'feebate' would provide that when a new building is connected to the electricity, water or sewage system, its owner must pay a fee for its inefficiency: the less efficient it is, the bigger the fee. But there's a corresponding opportunity for the owner to be rewarded instead. If the building is *more* efficient than normal, and therefore saves everyone money, then instead of incurring a fee it will win a rebate, and the more efficient it is, the bigger the rebate, without limit.

Best of all, year by year *the fees pay for the rebates* (and for the minor cost of administering the system). Read our lips: Not a new tax. This 'revenue-neutrality' is politically very attractive in these antitax days, and manifestly fair to everyone. The feebate simply transfers wealth from those whose inefficient choices impose social costs to those whose efficient choices increase social wealth.

The details are similarly simple. The 'slope' of the fee – how steeply it rises with inefficiency – should be rationally related to the long-run social cost of meeting increased demand for whatever the inefficiency is wasting. The slope of the rebate – how steeply it increases with efficiency – should more than cover the builder's cost of achieving the resource savings, so the builder or owner will make a profit up front. And, obviously, the area under the fee part of the curve should equal the area under the rebate part (plus transaction costs), changing the values each year so that this always occurs. Of course, under the influence of the feebate, buildings will become better each year. Eventually people will no longer put up inefficient buildings, so there won't be enough revenue from fees on inefficient buildings to pay rebates to efficient ones. At that point, one simply declares victory and stops.

Notice that, unlike standards, feebates are technologically dynamic. The greater the resource efficiencies that become possible and worthwhile, the bigger the savings people will be encouraged to buy (and discouraged to ignore). This happens automatically; no revision of standards is required. In fact, no standards are required either, although it's probably worth keeping them as a minimum-performance 'backstop'.

Compliance Tools

To be sure, someone has to assess how efficient a new building is in order to calculate what fee or rebate it merits, but that's not much different than enforcing standards right now. In fact, places like California have developed excellent tools to make it easy to comply with building standards that have two options – either 'prescriptive' (do exactly what the book lists as an acceptable package of measures) or 'performance' (keep total energy use below a specified level in any way you want).

To help builders comply with 'performance' standards without having to become experts in building physics or computer analysis, the California Energy Commission provides, for each of the state's 16 climatic zones, a free compliance handbook with helpful explanations and a simple table using a 'point system'. If the new house has so much area of this kind of window facing in that direction, you look it up in a table and find that it's worth so many points, which you write in the blank provided. A furnace is worth so many points depending on how big and how inefficient it is. Insulation, appliances, lights, each attract a certain number of points you can easily look up. Then you just add up all the points. For a typical family, in a year with average weather, the total number of points will quite accurately predict how much energy the house will use. A simple formula, table or graph can then predict how much money that energy will cost.

Once you know the house's energy score, it's a short step to saying that for houses built this year, houses scoring over x points will pay a fee (so many dollars per extra point), while houses scoring under x points will get a corresponding rebate. For that matter, a local government could require, or smart builders could volunteer, to put the number of points on the FOR SALE or FOR RENT sign – perhaps a numerical score, or perhaps in some vaguer way, such as five stars for an extremely efficient house, and so on as for the quality of hotels. Then the builder would not only see the up-front price signal of the fee or rebate but would also know that forever after (or until retrofitted) the house's market value would be affected by its energy score.

Feebates for Negalitres

The feebate concept has been tried or considered in half a dozen of the United States, or in localities within them, often successfully. A few political failures did occur, as in Maine, where fees were unwisely introduced before rebates, annoying builders of inefficient houses without gaining political protection from builders of efficient ones. Obviously, it would be smarter to introduce the system all at once, so everyone can see there's a fair opportunity for any builder to get rebates instead of paying fees; or, if the introduction is in steps, do the rebates first.

Something very like feebates has also been successfully used in some cases to encourage water efficiency. The architect of the Novi Hilton hotel in Novi, Michigan, requested and got a credit of $70,000 against the new 300-unit hotel's 'tap fee' to connect to the local water system because he'd specified water-efficient fixtures that he could prove would save 70 per cent of the water. He therefore didn't 'lock up' as much of the city's water and wastewater capacity as usual, leaving it available for other users and deferring the need for costly expansions. In addition, after 5 years' operation the hotel reported metered savings of $35,000–45,000 a year on water and sewerage bills, together with high satisfaction among its guests (Rocky Mountain Institute, 1994, pp 59–61).

Marin Municipal Water District in California charges a one-time fee of $13,840 for each acre-ft (1,234 cu m) of annual consumption for new hook-ups and for additions to existing hook-ups. But since the fee depends on estimated water use, not on land area, the efficient get smaller fees. For example, an acre of water-hungry lawn would incur a hook-up fee of $55,360; low-water-use plants, $20,760; and drip-irrigated low-water-use plants, only $6,920. At that rate, lovers of Kentucky bluegrass get to pay for their being so far from Kentucky's lavish rich rainfall, and those who prefer water-frugal but beautiful native plants save a bundle.

In North Marin County, California, builders of townhouses or condominia who limit their turf area per unit to 400 sq ft (or 20 per cent of the total landscaped area, whichever is less) get a $190 discount on the normal hook-up fee per unit. Turf area of 200 sq ft or less earns an apartment builder a discount of $95 per unit. These voluntary turf limits amounted to a 40 per cent reduction in previously typical turf areas, corresponding to reductions in total water use of about 16 per cent for townhouse units and 8 per cent for apartments. But in a cut-throat market, builders soon learned that few buyers wanted more lawn and costlier housing: in the programme's second and third years, over 95 per cent of new units qualified for the credits. The 'cash-for-grass' programme even extends to retrofits: the district pays a 50-cent rebate for each square foot of turf converted to water-saving plants (up to a total of $310 per single-family house – the same as the rebate for limiting a new home's turf to 800 sq ft).

FEEBATES FOR VEHICLES

Perhaps the first truly large-scale use of the feebate concept will be to encourage fuel-efficient cars. For the past decade or two, different tax regimes have, in effect, conducted a huge experiment to see how people behave under two- to fivefold differences in the price of motor fuel between North America and Western Europe or Japan. In general, heavily taxed petrol modestly decreased how much people drove their cars, but had a much weaker effect on how efficient a car they bought in the first place.

On reflection, that's not surprising. Most of the costs of driving a car are fixed, like debt service and registration, or depend only slightly on how much you drive, like maintenance and insurance. In the US, with the cheapest petrol in the industrial world, only about one-eighth of the cost of driving is fuel. Therefore any price signal in the petrol is diluted 7:1 by the other costs of owning and running the car. In addition, most people discount future fuel savings very heavily over a relatively short expected period of ownership. And in many countries, so many cars are owned, run or at least paid for by corporate employers, usually as a form of tax dodge, that much, even most, of urban driving is by people who aren't paying their own driving costs at all!

New American cars (not pickup trucks or vans) are slightly more fuel-efficient than, say, new German or Japanese cars, especially if adjusted for similar size, performance or features. This surprise is caused by government fuel-economy standards, which brought the American automakers to about the same place that their overseas counterparts reached through heavy fuel taxation. America's Corporate Average Fuel Economy standards ('CAFE') were indeed largely or wholly responsible for doubling the average fuel efficiency of new cars. But since that achievement, hostile US presidents or legislators kept the standards stagnant or even rolled them back; over half the continuing improvement in fuel economy vanished into ever-faster acceleration, so the average new car can now drive at nearly twice the speed limit; and further tightening of the standards to accommodate advancing technology has been doomed by political gridlock. (Automakers say the answer is higher petrol taxes; oil companies say it's tighter standards; and both are powerful enough to block action.) Moreover, the standards are not easy to enforce, have some details that are not always fair, and are often 'gamed' by manufacturers. Also, as we shall argue in the next chapter, the CAFE standards have not led to a significant reduction of per capita fuel consumption (see Figure 21, p 202); they seem to have signalled to American car owners that they can now afford to drive more miles.

Indeed, comparisons of international data by Lee Schipper of Lawrence Berkeley National Laboratory clearly show that although petrol prices have only a modest effect on the efficiency of cars purchased, they have a substantially larger effect on the amount of driving people choose to do.

Happily, feebates offer an alternative that makes for easier and fairer enforcement, offers potential profits as well as costs, is resistant to 'gaming', and keeps up with – even encourages and elicits – technological improvement. In 1990, for example, the California legislature, by an unheard-of 7 to 1 margin, approved the 'Drive Plus' feebate. A new car more efficient than normal would receive a rebate, and one less efficient than normal would pay a fee, depending on their fuel-economy ratings.* In addition, a separate feebate applied to smog-forming emissions. The law never went into effect because the outgoing governor, George Deukmejian, vetoed it after the legislative session (because Ford Motor Company opposed it, although General Motors was neutral); but it is ripe for revival, perhaps even on a national scale.

Many variations on the feebate theme are possible. If legislators had agreed, as some argued, that more efficient cars would probably be smaller, weaker and less safe, they could also have provided a feebate to reward safer and penalise less safe design and construction as measured by standard crash-test scores. If legislators didn't want to create an incentive to shrink cars, they could have calibrated the feebate in terms not of miles or emissions per litre of fuel but rather of those quantities per unit of interior volume. Perhaps, too, there could be an advantage in paying rebates not to buyers but to manufacturers, as with compact fluorescent lamps (Chapter 5, p 166).

ACCELERATED SCRAPPAGE

Feebates offer an even more intriguing prospect: they can be used to reward people for getting the least efficient and most polluting cars off the road fastest. Just add a provision that the rebate for an efficient new car will depend on the *difference* in efficiency between the new car you buy and the old car you scrap. In effect, society offers you more money to recycle your car than your dealer offers you to trade it in. Bring in a death certificate and collect your rebate. Or

---

* The ratings were actually expressed as grams of carbon released per mile of driving, for the legal reason that the federal government regulates miles-per-gallon standards and therefore pre-empts the individual states from doing so, but the federal government does not regulate CO_2 emissions, so the states, according to some legal opinions, may.

scrap your functioning car and don't replace it – bring in two bumpers and a wing, perhaps – and collect your 'negacar' bounty.

How big could these rewards be? Estimates of the social value of eliminating an inefficient car suggest that the rebate should be on the order of several hundred thousand dollars per gallon-per-mile improvement between the old and new car (several thousand dollars per litre per hundred kilometres). That way, if you trade up from an old banger to a hypercar, you could get a rebate of around $10,000–20,000 – about enough to buy the hypercar. To the extent that society's least efficient cars have trickled down to the poorest people,* this approach would be progressive, helping those people cash in on the social value of their inferior cars and use the cash to buy very clean and efficient new cars that they could then afford to fuel.

Substituting the best for the worst cars yields very rapid cleaning of our cities' air and reductions in oil consumption. Perhaps a fifth of the American car fleet produces three-fifths of its air pollution; those dirty cars would be the first to get traded up. Such 'accelerated-scrappage' incentives would encourage competition, reward automakers for bringing efficient cars to market and open a market niche into which to sell them. Feebates may even break the political logjam that has long trapped the US in a sterile debate over higher petrol taxes vs stricter fuel-efficiency standards – as though those were the only policy options and small, slow, incremental improvements were the only technical options.

Cars are not the only opportunity for fruitful feebates. Buses, heavy lorries, trains, ships, even aircraft seem suitable candidates too. Cash-short airlines are reluctant to buy doubled-efficiency, even in some cases tripled- or quadrupled-efficiency,** jet aircraft, so fuel and pollution savings (not to mention noise reductions and safety improvements) of great value to society are deferred, while aerospace workers are made redundant for lack of work. Would it not make more sense to accelerate the scrappage of the worst aircraft and replace them promptly with the best?

No doubt ingenuity can greatly expand the kinds and uses of feebates. But even the simplest forms offer a transparently attractive and effective way to get less of what we don't want and more of what we do want, using the one to help pay for the other. Like taxing pollution to pay for resource efficiency, that's a double benefit: it helps to correct both kinds of problems sooner. And far from imposing heavy costs and bothersome mandates on corporations, it creates for them delectable new market opportunities.

---

* This was once true in the US, but no longer: by now it is the more efficient 1980s cars that are mainly owned by poor people, while the least efficient cars tend to be owned by children of rich households.

** Chiefly in replacing the former Soviet fleet with modern Western aircraft.

Chapter 7
Ecological Tax Reform

7.1 Let Prices Speak

Chapters 4 to 6 are inspired by the experience of superiority of markets over bureaucracies. They are equally inspired by disappointment with the reality even in so-called market economies. Perverse incentive structures, bureaucratic obstacles and vested interest in the status quo make it immensely troublesome to reap the potential bonanza of the efficiency revolution which technologically is right in front of us.

To overcome all these obstacles we have offered quite a few ideas for making the efficiency revolution a truly profitable business. Although these ideas are all based on successful experience and are all rooted in the market philosophy, we cannot deny that progress has been slow in spreading them to other places and other sectors. Maybe some of these splendid ideas, such as feebates or integrated resource planning, require just too many dedicated and well-informed entrepreneurs, legislators and activists, or just too many bureaucrats to administer and control the schemes.

An entirely different objection to all the instruments mentioned in Chapters 5 and 6 comes from the ecological camp. They fear that efficiency gains won't do the trick. Efficiency may help to buy time but will not, by itself, produce a signal for a lasting reduction of resource consumption – not as long as resource prices keep telling a seductive fairy tale of unlimited availability. As a matter of fact, resource prices have been remarkably stable over the past 150 years. The share of expenses paid directly and indirectly for natural

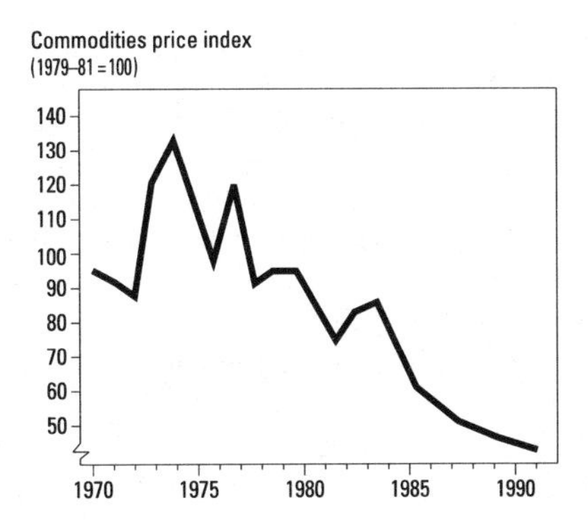

Figure 19 Resource prices fell drastically following the late 1970s despite a continuous, cumulative increase of resource consumption. The reason was aggressive and high-tech prospecting and mining after the oil shocks of 1973 and 1978. Markets tell little about the ultimate scarcity of resources or about environmental effects.

resources did not on average increase until the oil price shocks of 1973 and 1978. During the 1980s, however, commodity prices gradually returned essentially to their pre–1973 levels. Figure 19 shows that recent development. Only very recently, since 1994, have commodity prices begun to pick up once again, this time due to the increased demand from rapidly growing Asian economies.

As technological progress in prospecting and mining is likely to continue, market prices will keep telling our societies an untrue story of an ever fuller cornucopia of natural resources. When either resources or, more likely, the absorptive capacity of the environment are nearing exhaustion, and prices may start telling a new story of scarcity, it could already be too late. Even if we optimistically assume that the Factor Four revolution will dramatically reduce specific resource use, absolute consumption has no reason to go down. In other words, the Factor Four revolution would be used up and exhausted just for quadrupling consumption of goods and services, with nothing left for reducing overall resource use.

Yet as we are going to argue in Chapters 8, 9 and 10, the sustainability imperative from Rio de Janeiro requires drastic reductions in resource consumption. The successful creation of efficiency markets, namely the US utility reform (Chapter 5), was not actually intended to obtain, and has certainly not yet led to, a net reduction in total and overall energy consumption, although it has admirably increased specific energy efficiency and thereby served to halt the trend of further increases in resource consumption.*

Long-term scarcity and limits to the absorptive capacity of the environment are strong reasons for artificially raising resource prices. In the economists' language: 'external' costs should be internalised. The user of resources should pay the full costs, including costs to society, to the environment and to future generations. Of course, determining external costs is not an easy affair. Estimates by the Basle-based consulting firm Prognos (Masuhr *et al.*, 1995) for the German Ministry of Economic Affairs have produced figures of externalities for energy alone in the vicinity of 5 per cent of the German GDP. Barbir *et al.* (1990) go even further and suggest externalities as high as 14 per cent of (world) GDP for fossil-fuel burning alone. Similar figures of economic losses from environmental degradation are presented for developing countries in the Worldwatch Institute's 1995 *State of the World Report* (Brown *et al.*, 1995, p 12).

Artificial and predictable interventions are much easier to cope with than the unpredictable and inconsistent processes of world markets and world politics, such as the oil shocks which we saw in the seventies. Important business voices have recommended internalisation of external cost as a powerful tool for promoting eco-efficiency (DeAndraca and McCready, 1994, for the Business Council for Sustainable Development – since a 1995 merger renamed the World Business Council for Sustainable Development).

Internalising external costs should produce additional wealth. It should make countries richer, not poorer, according to textbook economics since the time of Arthur Cecil Pigou (1920). And what do we see if we look into the matter from an empirical angle? The Swiss economist Rudolf Rechsteiner (1993) has compared OECD countries' economic performance since the mid-seventies (*i.e.* since the time energy prices were very important) and found that their performance showed a positive, not a negative, correlation with energy prices in the respective countries (Figure 20). Japan fared best by concentrating on sophisticated production and letting the bulky, old-fashioned industries emigrate.

The former Communist countries, not included in Rechsteiner's graph for lack of reliable data, had energy prices far lower still, and performance far poorer, than all the OECD countries. Energy

* In California it has also modestly decreased electricity use per capita despite rapid economic growth. Indeed, that growth was substantially stimulated by efficiency investments' avoidance of thousands of millions of dollars' worth of unnecessary power stations, whose capital could then be reinvested more productively elsewhere.

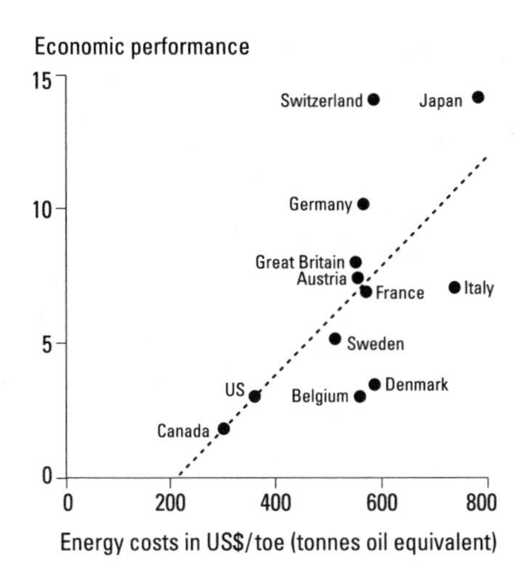

Figure 20 A positive, not negative, correlation can be found between economic performance (measured by an indicator score for growth, balance of trade, patents and the like) and average energy prices in OECD countries (from Rechsteiner, 1993).

prices in the Communist countries signalled to the energy user that energy was a free good, which led to unimaginably wasteful structures throughout those economies. Nevertheless, at the present state of analysis, we do not insist that Rechsteiner's correlation says more than that we need not necessarily be afraid of rising resource prices.

Another observation seems to support the use of direct price policies. Jochen Jesinghaus (Weizsäcker and Jesinghaus, 1992) established a striking negative correlation between fuel prices and per capita fuel consumption. Figure 21 shows his results. Ten years after the introduction in the US of the Corporate Average Fuel Economy (CAFE) standards, that country, although admirably catching up to Europe on per-mile fuel consumption, was still the country with by far the highest per capita fuel consumption. (In other words, under the condition of low fuel prices, what CAFE conveyed to automobilists was: 'Now you can drive more miles for your bucks.' Which they did.)

Careful statistical analysis shows that fuel prices are indeed by far the most important factor influencing per capita fuel consumption. Factors with rather marginal influence include average wealth and overall population density.

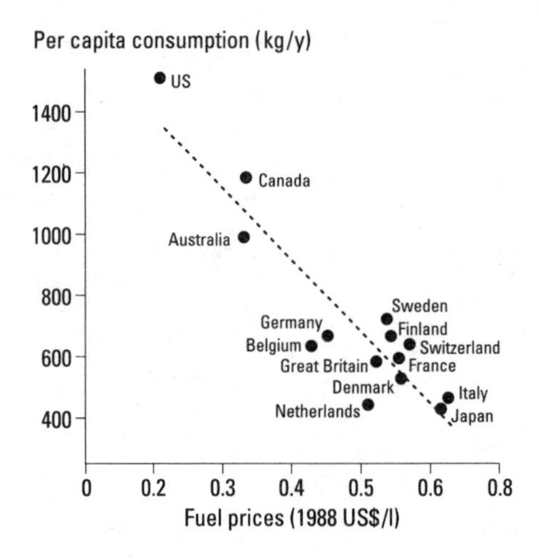

Figure 21 A strong negative correlation can be observed between fuel prices and per capita fuel consumption in countries of equivalent status of prosperity. The graph indicates a near-perfect, if long-term, price elasticity of fuel consumption. (From Weizsäcker and Jesinghaus, 1992)

Now we have four pieces of evidence:

- fuel prices correlate negatively with fuel consumption (we observe a high, if long-term, price elasticity);
- energy prices appear to correlate positively with economic performance, not negatively as conventional wisdom from the industrial lobby suggests: that is, cheap-energy economies tend to be wasteful and uncompetitive while dear-energy economies tend to be ingenious, innovative and highly competitive;
- higher resource prices are justified as a means of internalising external costs; and, best of all,
- a fourfold increase of resource productivity is technologically available and often cost-effective, so that just no loss of well-being must be feared from rising resource prices.

These four reasons should make politically feasible a proposal for strategically increasing resource prices.

Technically, resource prices could be increased by introducing tradeable resource quotas (the fewer the dearer) or by direct pricing. We remain completely open to both these approaches. But industry should be warned that the instrument preferred by mainstream economists, tradeable permits, can become quite brutal and unpredictable. Imagine what the North would have to pay to the South if carbon emissions permits were based on an equal per capita basis. And imagine how world markets for such permits would react to a 5-year period of 10 per cent annual growth in China. In the end, it may be reasons of familiarity and practicability that make direct pricing via taxes the preferred instrument. Appreciating, however, the extreme political sensitivity of taxes, we restrict our proposals to a strictly revenue-neutral tax reform. The state shouldn't become fat, but those players who exploit the opportunities inherent in the efficiency revolution should have the opportunity to satisfy considerable avarice.

7.2 The Least Bureaucratic, the Least Intrusive and Arguably the Most Powerful Instrument

Ecological tax reform (ETR) is actually a very old idea. Its roots trace back to the British economist Arthur Cecil Pigou (1920) who observed that it would be good for the economy if fair prices were paid for the consumption of common goods. Taxes, he felt, should help adjust prices accordingly. In our modern times, Pigou's classic book has often been quoted to justify taxes on the use of ecological goods.

One of the most exciting modern statements about ecological tax reform comes from Bob Repetto, Roger Dower and their colleagues of the Washington-based World Resources Institute. In their Green Fees study, they say that 'the total possible gain from shifting to environmental charges could easily be $0.45 to $0.80 per dollar of tax shifted from "goods" to "bads" with no loss of revenue' (Repetto *et al.*, 1992, p 11). They derive the statement from a learned paper by Charles Ballard and Steven Medema (1992) of Michigan State University who have been looking at the damage inflicted on our economies by our existing taxes, which can justly be labelled 'perverse taxes'. These plainly penalise the 'goods', namely human labour and the successful use of capital. In Europe they also penalise the successful creation of added value. Small wonder that reducing such perverse taxes would make the economy stronger, not weaker. But the state will need some revenues. The best way out is penalising the 'bads'.

ETR is not a simple or an unequivocal concept. Doubtless hundreds of ways exist of doing it badly and thereby damaging the economy and further reducing public confidence in the taxing authorities. After years of discussion, chiefly in Europe, we have come up with a basic proposal which may still be subject to modifications, depending on economic circumstances and the political climate. We propose a revenue-neutral, slowly progressing long-term tax shift. For a first approximation in quantitative terms, we say that end-user real prices for energy and primary resources should increase by around 5 per cent annually for a period at least of some 20 years, preferably 40 years or more. Emphasising end-user prices means that the widely differing price levels of today have to be acknowledged, *e.g.* between industry (ca. $2.5 per gigajoule), households (ca. $6 per gigajoule), car fuels (ca. $15 per gigajoule) and electricity (ca. $20–50 per gigajoule for different customers). Percentages other than 5 per cent are also conceivable; even a zero generous announcement phase (*e.g.* for industry) can be considered to allow time for adaptation. In any case, the annual signal should be so mild that no capital destruction would result and that technological progress in average resource productivity can outweigh the price increase, thus leaving constant the average annual expenditures for energy and resources.

Even if resource prices were increasing slightly more rapidly than efficiency, no harm should be expected from ETR. If average annual efficiency gains were 3 per cent while resource prices were increased by 5 per cent each year, then resource expenses would increase by a mere 2 per cent annually. For a family spending less than 5 per cent of its monthly budget on energy, that price increase would amount to less than one-tenth of a per cent of its annual budget. At the same time other treats, notably in much-needed personal services, would become more affordable as the fiscal drag on labour receded. So the signal of a well-designed ETR would be socially acceptable.

Nonetheless, the same signal would be tremendously strong for technology development. Knowing that energy and resource prices will steadily go up by 5 per cent per annum for a very long period of time would serve as an extremely powerful motivation for managers and engineers to work on the efficiency revolution. Suddenly you would find hundreds of businesspeople snooping around for bonanza opportunities in nearly all of the 50 arenas indicated in Part I of this book, and no doubt in many more as well.

Growing Support

ETR is gaining support from a wide variety of players. Economists are beginning to appreciate the potential in ETR for changing perverse incentive structures, for reducing undesirable taxes and for environmental deregulation. If prices begin to reflect ecological costs, products and firms damaging the environment would no longer have competitive advantages as is so often the case now. In Paul Hawken's words: 'Competition in the marketplace should not be between a company wasting the environment versus one that is trying to save it. Competition should be between companies which can do the best job in restoring and preserving the environment' (Hawken, 1994, p 90).

The next group of supporters of ETR are those who are concerned with high unemployment. For the European Commission, jobs are the top priority. In former Commission president Jacques Delors's *White Paper on Growth, Competitiveness and Employment* (1993), ecological taxes are depicted as a cornerstone for the Commission's fight against unemployment. For Delors's successor Jacques Santer and his Environment Commissioner Ritt Bjerregaard, this credo is just as valid.

Traffic planners, too, have discovered the elegance of green taxes. In Britain, with its notorious traffic congestions in the south of England, the Conservative government introduced a steadily increasing petrol tax, meant to increase by 5 per cent annually over an undetermined number of years.

Environmentalists need hardly be mentioned among the supporters of ETR, although their enthusiasm for green taxes is surprisingly new. In the beginning, their favourite approach was command-and-control, as they feared in ETR 'the rich would buy themselves out of their ecological obligations whilst the poor would have to adapt.' Meanwhile, the ecological camp understood that the price signal is working just as well on the 'rich', mostly via the technological revolution it can elicit.

Also, enlightened businesspeople like Stephan Schmidheiny from Switzerland, who created and chaired the Business Council for Sustainable Development, are supporters of ETR. Schmidheiny's concept of 'eco-efficiency' (Schmidheiny, 1992) resembles the philosophy elaborated in this book. For the business community, the charms of ETR also relate to its potential for much further deregulation. If prices begin to tell the ecological truth, why employ thousands of bureaucrats and lawyers to say the same in a much more complicated language? ETR may one day be seen as a strategic method for creating 'nega-bureaucrats'. ETR can be seen as the least bureaucratic and the least intrusive method of steering the economy in the desired direction.

The next major group of supporters for ETR will be the engineers worldwide who see the technological scope of the efficiency revolution and are impatiently waiting for the new continent to be discovered. Provided the signal is reliable enough, capital would massively flow into R&D for hundreds and thousands of new resource efficiency technologies that are as yet barely conceived.

The economic justification for ETR along the lines of Arthur Pigou comes from considerations of external costs. If the estimates quoted above about externalities are reliable, green taxes reaching as high as 10 per cent of GDP would be justified and should be expected to improve the economic performance of the country. For the US economy, 5 per cent of GDP is some $300 thousand million! Even above that level, a revenue-neutral ETR could lead to economic gains if the losses inflicted by previous taxes were higher than potential losses resulting from the new tax. But this book is not meant to provide elaborated proposals or an extensive discussion of ETR. That has been done elsewhere (Weizsäcker, 1994, chapter 11; Weizsäcker and Jesinghaus, 1992; Greenpeace, 1995).

One idea is a nondiscriminatory tariff on 'grey energy', *i.e.* energy incorporated in materials such as aluminium, steel or chlorine. Being non-discriminatory (not disfavouring foreign competitors), the tariff should not pose a major WTO (World Trade Organization) problem (Westin, 1997). Yet it would 'protect' domestic manufacturers against those from abroad whose competitive edge consists only of cheap energy inputs. Similar ideas presently being discussed in Brussels include the introduction of an Energy Added Tax (EAT) similar to the existing VAT, and a high VAT level for energy and raw materials.

7.3 MUCH SCOPE FOR INTERNATIONAL HARMONISATION

ETR should ultimately receive strong support from those in the business community and elsewhere who have to think internationally. It is always difficult for firms in the US or in Europe to adopt high environmental standards that may be ignored by their competitors abroad. To a certain extent, such standards are being adopted anyway with a view to maintaining a favourable image with clients and also to keep abreast of new developments. Also, as has been indicated above, some protective measures can be taken without questioning the GATT/WTO (World Trade Organisation) principles of a free world market. But everyone, both in the business world and in the environmental camp, would feel greatly relieved if standards and rules could be harmonised internationally.

ETR may be the most promising strategy for doing exactly this. If a country becomes richer, not poorer, by applying ETR, then why should even a poor country hesitate to introduce it? The first step in any case should be desubsidising resource use. Who could possibly contradict this (except those pleading to keep their own subsidies)?

The 'nega-bureaucrats' argument could prove particularly powerful. Estimate the administrative costs in Brazil, Zaïre or India for collecting a million tax dollars from dentists, lumber traders or businessmen maintaining accounts in the Cayman Islands. Compare these costs with those for collecting a million tax dollars from a coal mine or an oil tanker unloading at any petroleum port. Further estimate the administrative costs in those countries for establishing a command-and-control regime of environmental protection and compare them to the costs of environmental tax collection.

The social equity objection is, of course, that one must not collect taxes from the poor who are dependent on energy use in order to spare the income of the rich. Well, that's neither the intention nor the effect. In the developing countries (unlike in OECD countries) the rich are disproportionate consumers of energy. They are the ones using aeroplanes (often private ones), owning a car fleet, air-conditioning their villas and offices and eating food that has travelled great distances. If noncommercial renewable energy sources such as biogas, biomass and small-scale solar, water and windpower are freed from the tax, and a small tax-free subsistence allowance is granted for metered electricity and small-scale oil use, no social problem should arise.

The strongest argument, however, for introducing ETR in developing countries, especially in the rapidly industrialising countries of Asia and Latin America, has more to do with the central message of this book. For this group of countries, the efficiency revolution is even more urgently needed than for the old industrial countries of the OECD.

The US or Japan or France is theoretically rich enough just to buy and waste a lot of energy and materials. It is highly unreasonable but not a disaster. But for India, Egypt or Colombia, it's tragic. Capital is scarce, energy is scarcer, labour is abundant. Why should these countries concentrate capital spending (even if it is soft-loan money from the World Bank) in aggressively expanding resource exploitation while ignoring the opportunities shown by the Factor Four revolution? Why should they emulate 20th century robotics and neglect the efficiency technologies that are likely to characterise 21st century technology markets and that are so much closer to the actual needs of their countries?

Whatever the World Resources Institute has said in the US context about gains rather than losses from ETR should apply even more to China or India, where penalising capital or labour and underpricing scarce energy is even more dangerous than in the rich countries.

Some countries will, of course, object to ETR. The OPEC countries are clearly opposed to it. If the world wants to put an additional price tag on oil and gas, they argue, then that money should go to them, the producers. They have a point there. Some compromises may be negotiated so as to foster international understanding and a feeling of mutual trust and equity. But ultimately, levying domestic taxes on undesired commodities (or on pollutants associated with the commodities) is something sovereign states can decide for themselves.

It will be essential for the commodity-exporting countries to recognise that the age of resource scarcity they all have dreamt of in the seventies may never become a reality. Progress in efficiency is likely to speed up and ultimately to exceed all potential market expansion for commodities. It will be wiser for these countries to go with the trend and to become engaged in the efficiency revolution. And the World Bank and other lending institutions should swiftly stop most of those destructive schemes of further resource exploitation which tend to depress world market prices for the respective commodities. That mechanism benefits the North more than the South to whom the benefits allegedly are meant to accrue.

In Part III of this book we will introduce strong additional arguments in favour of the efficiency revolution and for ETR. It is the message from Rio de Janeiro, sustainable development, which will very likely induce the world to stop overexploiting resources and underusing human labour. Who out there really wants to miss the boat? The race is on between the North and the South, and within the North, for the competitive edge in resource efficiency. And ETR may emerge as the favourite instrument both in the North and the South for making the race a smooth and humane transition.

The Scandinavian countries and The Netherlands have already

started with ETR. Notably Denmark has established a CO_2/energy tax exempting industrial process energy, fully compatible with industry's needs. The Dutch model, introduced in 1996, offers an exemption from the tax for the basic energy needs of poorer households.Belgium, Germany, Austria and others have introduced green charges on certain products. The UK has introduced the petrol tax escalator and a landfill tax. Germany has introduced many detailed charges and tax differentials intending to create an incentive for ecologically benign developments (but has been reluctant to introduce any form of an energy tax). The European Union will in any case pursue and support the idea. The time seems to have come for some international agreement.

Part III
A Sense of Urgency

We introduced the efficiency revolution in the first half of this book. It is going to be profitable for its pioneers, a little less so for their followers. Unlike classical pollution control, efficiency will not be a question of sacrifice. Nevertheless, there are also very strong ecological reasons behind the efficiency revolution. Reading any of the recent *State of the World* reports by the Worldwatch Institute (*e.g.*, Brown, 1995) or reflecting on *Beyond the Limits* (Meadows *et al.*, 1992) will suffice to convince us that humanity is on a collision course with natural boundaries. If we fail to change course soon enough and the collision occurs, nature will survive the event *somehow.* Humanity will not.

We now undertake to outline the nature of the global environmental crisis. We make use of the earlier Reports to the Club of Rome including *The First Global Revolution* (King and Schneider, 1991). Also, we relate our diagnosis to the results from the 1992 Earth Summit in Rio de Janeiro, the biggest diplomatic event so far to address environmental issues.

In Chapter 8, we introduce the three big urgent themes of this historic summit: sustainable development in general, climate and biodiversity. And we add some observations about some other unsolved ecological problems. In this and the subsequent chapter, we shall see that Factor Four will make a huge contribution towards answering the Rio challenges.

One observation that goes beyond the talk of Rio de Janeiro will be that we just can no longer afford to 'ignore the megatonnes while shooting at the nanograms.' The 'forgotten agenda' of material flows will be discussed in Chapter 9.

Compared with Factor Four solutions, classical environmental policy will look unsatisfactory, risky and costly, as we shall argue in Chapter 10.

Finally, we present the Factor Four efficiency revolution as the most powerful strategy for closing those abyssal gaps that are opening before us. The most modest ambition in this context is to *buy time* during the unavoidable phase of human overpopulation of the planet.

Chapter 8
The Challenge from Rio

8.1 The Earth Summit and *The First Global Revolution*

The Earth Summit of June 1992 in Rio de Janeiro, officially called the United Nations Conference on Environment and Development (UNCED), has been celebrated as the biggest gathering of heads of state and of government in all human history. More than a hundred top representatives of states came, not counting the hundreds of ministers who also attended. In the end 30,000 people made it to Rio, most of them spending all their time at the multitude of special events held at the Global Forum, which was arranged some 20 km away from the conference site.

Two major conventions were adopted at UNCED, the Climate Framework Convention and the Biodiversity Convention. The protection both of the global climate and of global biodiversity had emerged during the preceding decade as the first priority of world-wide environmental policy (see Chapter 8, p 222–230). Fortunately for the delegates, the conventions had been negotiated beforehand so that the heads of state and of government could sign them without further ado.

Most of the conference time was needed to finally agree on each paragraph of *Agenda 21*, the 40-chapter master plan for a 21st century policy as well as on certain final formulations of the 'Rio Declaration'. Official speeches relating to Agenda 21 and the Declaration invariably stressed the need to protect the environment. However, all Southern delegates emphasised the need for further economic development.

The two principles of environment and development were encapsulated in Principles 3 and 4 of the Rio Declaration as follows:

> *(Principle 3) The right to development must be fulfilled so as to equitably meet developmental and environmental needs of present and future generations.*

> *(Principle 4) In order to achieve sustainable development, environmental protection shall constitute an integral part of the development process and cannot be considered in isolation from it.*

Many Northern delegates opposed the recognition of a right to development in the context of the Rio Declaration. They argued that if such a right existed, it was a limited right, constrained by natural boundaries, *i.e.* the limits of natural resources and the capacity of the ecosystem to restore itself. When the inclusion in the Declaration of a right to development became diplomatically unavoidable, Northern delegates suggested at least combining Principles 3 and 4 to indicate some conditionality. But that is precisely what was not acceptable to the developing countries. They explicitly prevented any downgrading of the right to development in Principle 3, as they saw it, to the mere right to sustainable development.

Where does this fear of qualifying development come from? Well, that's not so difficult to understand. To begin with, when the North developed, nobody cared about limits to natural resources or the ecosystem's ability to restore itself; and whenever some resources seemed to be getting scarce, expeditions would go to the 'colonies' and dig and harvest and ship back to the North whatever was needed.

That was, of course, before modern environmental policy came onstage. But from a Third World perspective, those modern pollution control policies were in a sense no better than the old-style resource exploitation. The developed countries kept saying that pollution control was very costly and therefore was affordable only in times of a robust and prosperous economy. That prosperous economy, on the other hand, was and is characterised by per capita consumption rates of natural resources easily 5 times, and in many cases 20 times, higher than in the developing countries.

Hence we from the North kept sending the message to the world that development (meaning high per capita resource consumption) logically and chronologically came before environmental protection. In effect, at home the North treated sustainable development as if it were dependent on a prior phase of unsustainable development.

The North has to understand that sustainable development worldwide simply will not happen unless and until the North itself learns to live with far smaller *per capita* rates of resource consumption. This is why we see Factor Four (in the North) as a target for and a prerequisite of sustainable development.

The First Global Revolution

The Earth Summit and its debates about sustainability have shown the world that there is no longer any way of separating environment and development. But the interconnectedness of issues goes much further. The Club of Rome in its first-ever Report *by* the Club (after 20-odd Reports *to* the Club) made this interconnectedness of what the Club dubbed the world problematique a central political theme. Authored by Alexander King, past president of the Club, and Bertrand Schneider, its secretary general, the report was called *The First Global Revolution* (1991) – quite a sweeping concept, but justified. The book outlined in a compelling manner how at least ten mutually interconnected problems constitute the world problematique:

- armaments and armed conflicts;
- the scandalous economic gap between North and South;
- population increase and food shortage;
- environmental degradation, growing energy demand and the greenhouse effect
- the urban megalopolis trend, chiefly in developing countries;
- the collapse of socialism, which did not solve local and ethnic problems in the area, notably in the former USSR;
- the economic tensions and cultural differences in the Triad (US, Japan, Europe);
- the widespread prevalence of emotional misery;
- the manifold and new problems of the information society; and
- the general governability problem both at the national scale in modern democracies and, more alarmingly in the context of the world problematique, at the global level.

This enumeration alone is enough to make you shiver. If the Earth Summit had difficulties coping with just the environment-development nexus, which international body or system will ever be able to handle that ten-sided nexus? The Club presents some hope by boldly moving from the problematique to the 'resolutique', a bundle of priority actions for tackling the problem. The list includes:

- conversion from military to civilian production (the authors wisely, and contrary to the widespread and quack notion of peace dividends, warn of substantial costs to be shouldered in the early part of the process);
- new environmental policies with a strong emphasis on a massive worldwide campaign for energy efficiency (we could not agree more!);
- new initiatives for a development of the South, including population control initiatives, and with a strong emphasis on rural development;
- governance taken seriously, emphasising consensus orientation and the international dimension;
- a systematic use of education and the media for the necessary transformation; and finally
- a wisdom- and solidarity-oriented change of global consciousness.

All this is fair enough. But in the real lives of politicians, there are still elections to be won under conditions of economic egotism, of nationalism, provincialism, fundamentalism and other -isms; and there are always political rivals making their careers out of appealing to any of those -isms. Much as we admire the comprehensiveness of the initiatives proposed by the Club, we feel that we should contribute to the 'resolutique' one project of a realistic agenda which does not have to wait until better ethics and world consciousness have transformed the leaders and their electorates.

8.2 Sustainable Development Is Inescapable but Has Hardly Begun

Sustainable development was not invented by the diplomats at the Earth Summit. Sustainability has been a guideline for human culture from time immemorial. Even animal populations had to observe it long before humans lived on earth. Parasites and predators have to be careful not to exterminate those upon which they live. The 'struggle for survival' according to Charles Darwin was more a struggle for the preservation of scarce resources than a battle of voracious predators against their prey. Overvoracious species are unfit in the Darwinian sense!

Humankind has for most of its history lived with a tacit understanding of the rules of sustainability. It was largely unnecessary to have an explicit consciousness of the concept, because resource use and population growth remained sustainable without much active intervention. But there were exceptions. Perhaps the most visible examples of unsustainable resource use relate to forests and to fish stocks.

Concerning forests, Germans often claim that their forebears invented the concept of sustainable yield. Forests were disappearing in Central Europe at alarming rates until the first decades of the 19th century. When coal was discovered as a readily available fuel, the need for burning wood receded, despite the growing energy needs of early industry, and made it possible for the German states (the kingdoms or dukedoms of Prussia, Bavaria, Oldenburg, etc) to establish regimes of sustainable harvesting for their mostly state-owned forests. Although, judged from today's perspective, the ecological value of the emerging fir and spruce monocultures was very doubtful, at least a sense of sustainability was established in the German culture. This ethos can easily be called upon when there is a need to pull the Germans onto the boat of sustainable development.

To be fair to history, however, the German claim is incorrect in two major regards. Firstly, many cultures before 19th century Germany had a much broader and more ecologically sound understanding of sustainable yield. Especially the Native Americans, both in South and North America, had and still preserve a cultural understanding of living in nature without mining it. The Native North Americans were shocked by the white settlers' wanton massacre of the prairie buffalo. And the so-called speech by Chief Seattle (formulated later by unknown authors) expressed the understanding that the very fact of owning and selling land for private exploitation contradicted the native people's religion: it was at first incomprehensible and then, when comprehended, blasphemous.

Second, the German sustainable forestry concept (stemming from earlier centuries) was enforced only after the spreading of unsustainable coal use. Hence, it was and still is a 'parasitic' concept. If coal, gas and oil had not been available to central European industries, the exploitation of forests would no doubt have continued domestically or abroad with no regard for sustainability.

We are sharing this history of 'sustainable forestry' because it may warn readers against fallacious concepts in the context of modern sustainability quackery.

The contemporary discussion of sustainable development has its origin in the World Conservation Union's study (IUCN, 1981) of sustainable resource use. From there the concept made its way into the World Commission for Environment and Development, the Brundtland Commission. That report (WCED, 1987) made sustainable development the cornerstone of its attempt to reconcile developmental and ecological objectives. The commission accepted the still somewhat imprecise definition: 'Humanity has the ability to make development sustainable – to ensure that it meets the needs of the present without compromising the ability of future generations to meet their own needs' (WCED, 1987, p 8).

Northern Lifestyles Were not up for Negotiation at the Earth Summit

This magic formulation does not, however, resolve the conflict. One facet of the conflict, at least, remains: Who is addressed by 'sustainable development'? The North keeps believing that the concept of sustainable development chiefly addresses the task of ecologising the South. The South, by contrast, feels that the term refers to the unsustainable lifestyles of the North. The Rio Declaration allows both interpretations. While Principles 3 and 4 quoted above implicitly talk about the South (although making cautious use of the term 'sustainable'), Principle 8 also implicitly speaks about the North:

> *(Principle 8) To achieve sustainable development and a higher quality of life for all people, States should reduce and eliminate unsustainable patterns of production and consumption and promote appropriate demographic policies.*

The reference to appropriate demographic policies is aimed at the South. But, quite correctly, a nexus is made in this principle between unsustainable (per capita) consumption and population policy. With per capita consumption rates about 15 times higher in Germany than in India, the total ecological burden caused by 80 million Germans is likely to be higher than that of 900 million Indians. Figure 22 shows that 1,000 Germans consume roughly ten

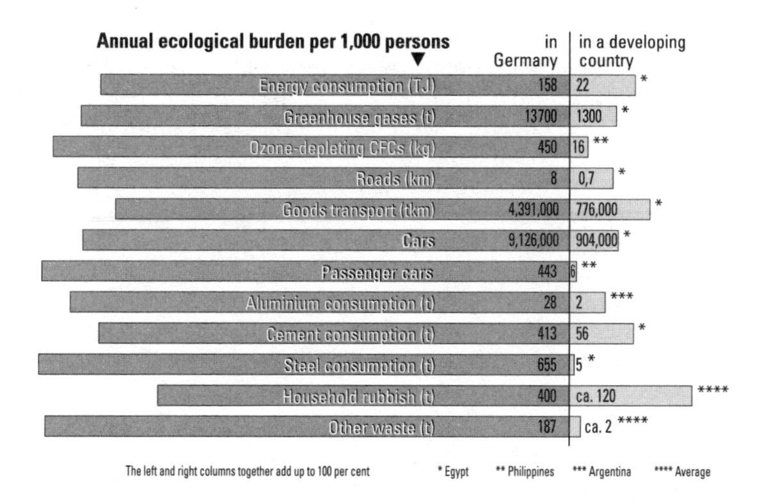

Figure 22 One thousand Germans consume roughly ten times more resources than 1,000 Argentineans, Egyptians or Filipinos. (After Bleischwitz and Schütz, 1992)

times more in various regards than 1,000 Filipinos, Egyptians or Argentineans. The graph was produced by the Wuppertal Institute in preparation for the Earth Summit and was widely circulated and quoted in the European media.

In the introductory remarks to this chapter, we associated ourselves with the view of the South. We feel that *per capita* calculations are central for assessing the sustainability of lifestyles and of civilisations. But we have to recognise that, at the Earth Summit, the North managed to get its view made official in that Agenda 21, the centrepiece of the UNCED negotiations, is mostly addressing questions relating to the South without in any serious way questioning the Northern way of life. The South was actually not unhappy with this fact, because Agenda 21 (other than the Rio Declaration) is not talking about principles but about practical things to do, with a lot of money meant to flow from North to South. Implementing all the proposals included in Agenda 21 would require the North to transfer a sum to the South worth roughly $100 thousand million annually. This sum is equivalent, intentionally or not, to 0.7 per cent of the accumulated Northern GDPs, the figure pledged by most Northern countries (though never by the US) as their annual development aid.

However, let us not deceive ourselves ecologically. If by any miracle that flow of aid money were ever reached, the global environment would hardly benefit. This is because the measures of Agenda 21 taken together would inevitably lead to an immense increase in construction, land use, energy consumption, traffic and, consequently, mining and forest destruction. It can be said, in defence of the agenda, that no other roads of civilisational development and no other methods of environmental protection than the Northern ones have so far been available. And any attempt at a fundamental questioning of the Northern model would have been strongly refuted by the Northern delegations. 'The American way of life is not on the table for negotiation,' said President George Bush when leaving for Rio de Janeiro.

FOOTPRINTS AND ECOLOGICAL SPACE: HOW CAN WE MEASURE SUSTAINABILITY?

In the context of limited planetary resources and ecosystem renewability, Figure 22 would seem to indicate that the Northern development model is fundamentally unsustainable. But what can we do about it? And what can we do when national development plans in China, Indonesia, Brazil, Nigeria and other developing countries pay less than lip service to sustainable development? If you challenge developing countries on this fact, they are quick to reply that the North did not fare too badly in rapaciously pursuing its own industrialisation – an approach to which they cannot see any realistic alternative.

There have been several attempts by researchers in the North to determine in quantitative terms what is unsustainable about the Northern model. There is William Rees and his team at the University of British Columbia, Canada, who have calculated that the ecological footprints of average Canadians are so large that we might need three globes the size of the earth to survive 5 or 6 thousand million footprints of that size; *i.e.* if all of humanity consumed and polluted at the Canadian rate, it would take three globes to accommodate us all (Rees and Wackernagel, 1994).

Similarly, a Dutch team under Manus van Brakel and Maria Buitenkamp (Buitenkamp et al, 1992) has estimated the ecological space claimed by average Dutch citizens, with essentially the same result. Their 'Sustainable Netherlands' study states that one overseas air journey, for example, would eat up your sustainable share of traffic energy for 3 or 4 years. Based on the same ecological space philosophy, the Wuppertal Institute has produced a report (BUND and Misereor, 1996) offering a number of 'Leitbilder'

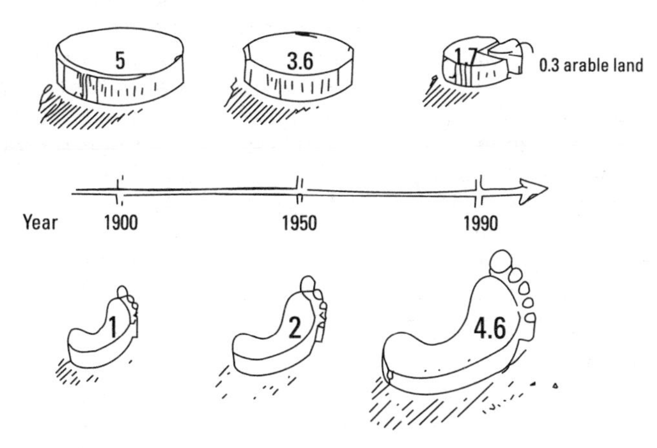

Figure 23 We need three earth globes to accommodate 6 thousand million 'ecological footprints' the size of typical Canadian ones. (Redrawn after Rees and Wackernagel, 1994)

(model concepts) that may contribute to new and sustainable lifestyles. Significantly, the analogous work in the US, published by the President's Council for Sustainable Development (PCSD, 1996) – though a major achievement of political consensus-building under the very unfavourable conditions of a Congressional majority that hardly cares about sustainable development – does not contain any mention of per capita limitations to resource use.

The impact in the real world studies suggesting per capita limits to consumption is, of course, very small. What conceivable government in the North would dare to ration air miles, auto miles, heating oil or water for your kitchen and bathroom? As long as our Northern ecological footprints remain as large as they are and, in fact, keep growing, we have no right and not the slightest chance to prevent the Chinese and all the others from following our deadly path.

It is convenient for the North simply to deny that the path is deadly. So the North keeps waiting to see if and when there will be any market signal inducing us to reduce petrol or water consumption – a wait-and-see attitude that is itself unsustainable. This may be bad news to many readers.

The good news is that highly attractive pathways now exist that would allow us to escape the dilemma outlined above. What can reduce the size of our footprints by a factor of four or more – and still offer us an equivalent to the American way of life in daily enjoyments – is the efficiency revolution. Part I of our book has shown 50 examples of how that revolution might work. Making all this happen now, with the help of the tools in Part II, is perhaps the easiest strategy for achieving sustainable development.

8.3 The Greenhouse Effect and the Climate Convention

The greenhouse effect is catching the imagination of people worldwide. Everybody is in some way dependent on the weather and the climate. The very idea that humankind is meddling with the weather is disturbing. Adding to the sense of unease is the knowledge that the world's well-to-do minority is producing the biggest contributions to the greenhouse effect. Quite uncommon coalitions have emerged among the 170-odd countries of the world with regard to climate protection, as we shall discuss in a moment.

Bangladesh is one of the poorest and most densely populated countries of the world. In the case of major climatic changes of the kind feared for the next century, Bangladesh, like the 30-odd countries of the Alliance of Small Island States (AOSIS), will be hit badly. Figure 24 shows the expected effects. A mere 20 per cent of Bangladesh's surface would not be in danger (Figure 24).

A rise in sea level is not the only climatic threat to the AOSIS states and Bangladesh. The incidence and strength of hurricanes is bound to increase as temperatures rise. The meteorological conditions for the creation of hurricanes depend upon surface water temperatures above 26°C. The areas in which these conditions regularly obtain are increasing.

All this worries not only the people living in these areas. Insurance companies are also very concerned. Just before the March 1995 Berlin Conference of Contracting Parties of the Climate Convention, the major European insurance companies got together to discuss the impact of changing weather conditions on their business. Media attention was caught by one illustration in particular, showing the drastic increase of storm-related economic damages during the last three decades (Figure 25). Hurricane Andrew alone, in August 1992, drove six American insurance companies into bankruptcy.

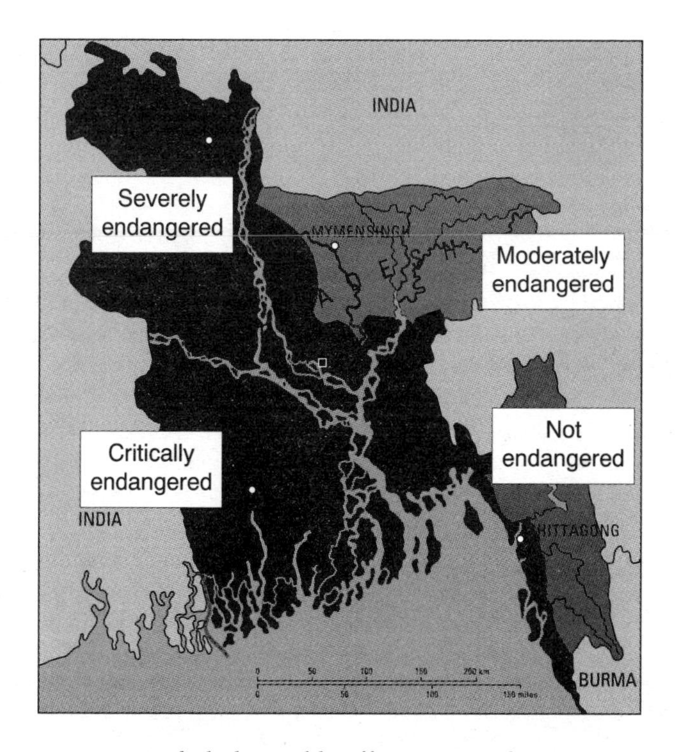

Figure 24 Bangladesh would suffer immense losses in case of climatic change. The areas coloured black, grey and light grey would be critically, severely and moderately endangered, respectively, if climatic changes occur as climatologists fear. (Source: Bangladesh Centre for Advanced Studies, 1994)

Is There Any Scientific Evidence?

The fear of a perilous greenhouse effect is actually not new. It was first uttered by Svante Arrhenius (1859–1927), the great Swedish physicist and chemist, in a learned paper published in 1896. He applied the atmospheric physics and chemistry of his day to contemporary experience of industrial carbon burning and concluded that humankind could easily effect a radical change in global weather conditions. A doubling of CO_2, he calculated, would lead to an average increase of global temperatures by 4–6C°. (A few years later, he said that the estimate was perhaps too high.) His scientific misfortune was that contemporary measurement methods were unable to establish the increase of CO_2 concentra-

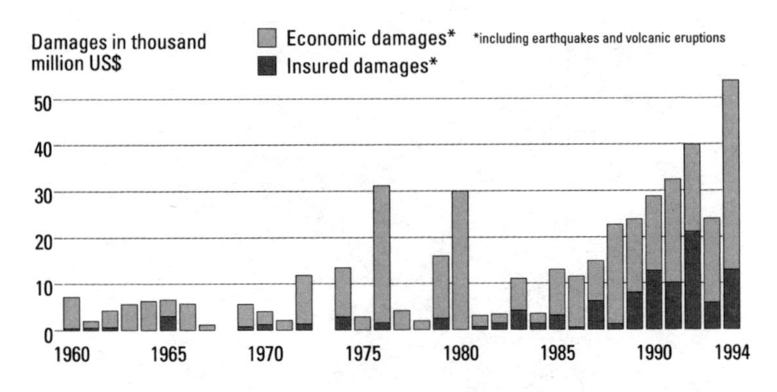

Figure 25 Increase of (mostly) storm-related damages worldwide, in thousand millions of dollars. (From Munich Re, after Der Spiegel, *1995)*

tions with anything like the precision required to prove his theory.

It was not until after World War II that continuous measurements of CO_2 concentrations in the atmosphere were available. The Hawaiian observatory on Mount Mauna Loa, located at a place with no interference from local emissions (which would make the measured concentrations dependent on wind direction and industrial activity) produced a striking upward curve of CO_2 concentrations over the years (Figure 26). The curve shows a trend of acceleration which corresponds to steadily increasing emissions. Scientists in the 1960s began to wonder where the trend would lead.

However, the Mauna Loa data were not sufficient to prove Arrhenius's theory. The correlation with temperatures was missing. The time-span was simply too short to significantly establish a corresponding trend of global warming. Slightly warmer summers and winters were observed but could still be attributed to statistical noise, to sunspot activity or to the long-term dynamics of interglacial climate change that went on without any human interference. Not surprisingly, therefore, the discussion of human-made climatic change didn't really take off in the 1960s and was soon overshadowed by more immediate ecological concerns, such as pollution of air and water. The agenda of the 1972 United Nations Conference on the Human Environment, held in Stockholm, concentrated almost exclusively on pollution control and made no mention of global warming.

A year later even these environmental concerns had to give way to something felt still more urgent: the 'energy crisis' resulting from

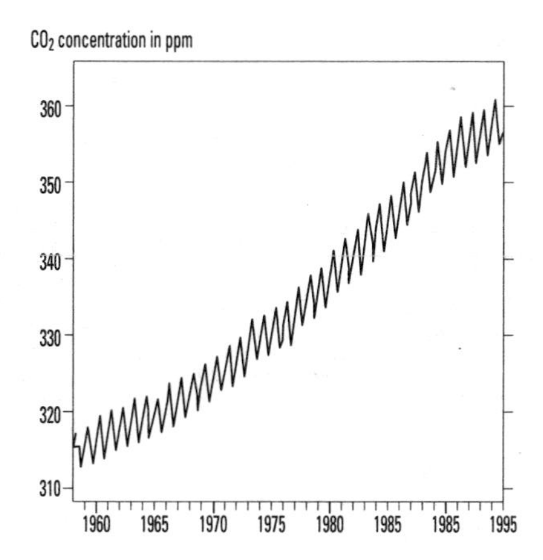

Figure 26 Carbon dioxide concentrations as observed far from industrial activity on Mount Mauna Loa, Hawaii, since World War II. The annual periodicity reflects plant growth (and thus CO_2 absorption) in summer in the Northern Hemisphere. The striking and systematic upward trend has led to the first serious reflections about a manmade greenhouse effect.

exploding oil prices that were dictated by OPEC. Energy concerns dominated the rest of the 1970s. Whoever in scientific circles was working on the Arrhenius theory could not realistically expect their subject to resonate with the public. The early 1980s were even worse in this regard. They began with the Soviet invasion in Afghanistan, the hostage crisis in Iran and the resulting cold war mind-set in America. This mind-set left very little room for environmental policy at all.

An unexpected scientific discovery changed all that. It was the record of 'fossil' CO_2 concentrations in the Antarctic ice. A deep drilling experiment launched from the Soviet Antarctic Vostok station produced a series of breathtaking data establishing an ongoing record of CO_2 concentrations for the past 160,000 years. A microchemical method developed by the Swiss climatologist Paul Oeschger allowed the French team of Claude Lorius (Lorius *et al.*, 1985; Barnola *et al.*, 1987) to analyse the chemical composition of tiny air bubbles enclosed in ice layers whose age was well known.

A more sophisticated method also allowed them to establish the average temperatures of the corresponding times. Oxygen consists of 99.8 per cent ordinary ^{16}O and 0.2 per cent 'heavy' ^{18}O. Water (H_2O) molecules loaded with heavy oxygen don't evaporate

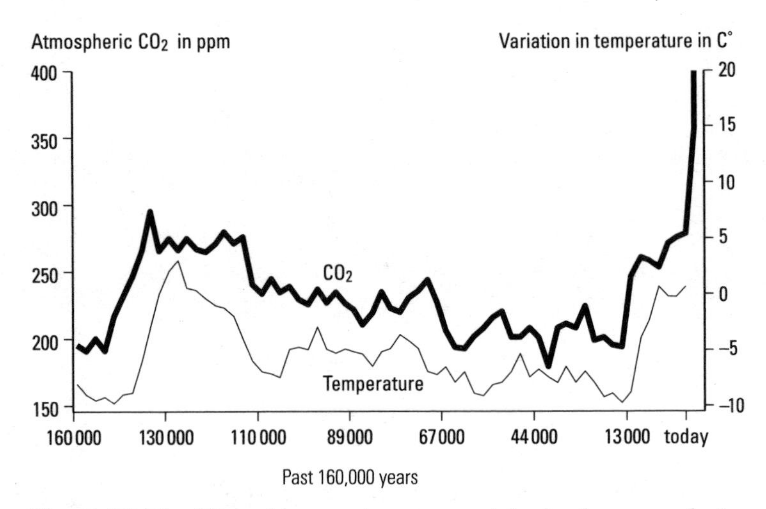

Figure 27 The 'Vostok' sensation: a surprisingly close correlation during the past 160,000 years between CO_2 concentrations and average temperatures on earth as established by the chemical analysis of 'fossil air' enclosed in Antarctic ice. (Drawing by Global Commons Institute, London, after Jouzel et al., *1987)*

quite as easily as those loaded with the ordinary oxygen atoms. Hence, water in the clouds contains a higher fraction of ordinary oxygen than does water in the oceans. Precipitation in Antarctica reflects the composition of water from the clouds, not from the oceans. Warm periods make for higher precipitation. Thus, ice stemming from warmer periods should contain lower concentrations of ^{18}O. This slight difference permits scientists to differentiate warm periods from cold periods well enough to establish a temperature record from the same air bubbles enclosed in Antarctic ice (Jouzel *et al.*, 1987). When the two curves of CO_2 concentrations and temperature were put together, a sensational similarity was found (Figure 27). It made the world think again about the dangers of an additional greenhouse effect caused by human activity.

When the Vostok ice core results reached the scientific community (before their publication in *Nature*), a cry of alarm was raised. The Geneva-based World Meteorological Organisation (WMO), thus far an inconspicuous UN organisation, the United Nations Environment Programme (UNEP) in Nairobi and the International Council of Scientific Unions (ICSU) in Paris got together and arranged a seminar in Villach, Austria, which was to become one of the most important environmental gatherings of our time. At the

Villach meeting, the evidence from the Antarctic ice core drillings was presented, and its political consequences were discussed.

It has to be admitted, however, that in much earlier geological times (500 million years ago and earlier) high temperatures coexisted with low CO_2 concentrations. According to the Arrhenius theory of global warming, this earlier pattern would not be based on direct causal relations. Rather, the greenhouse effect would have been masked by some other effects. Caution will in any case have to be applied in using the Vostok findings as the main basis for modelling climate and its future change.

A New Branch of Policy is Born

The Villach meeting can be seen as the hour of birth for a new branch of policy, climate policy. Hectic international and national activities ensued. The Second World Climate Conference was soon convened in Geneva, in 1988. To guarantee continuous governmental attention to the problem, the Intergovernmental Panel on Climate Change (IPCC) was established, which brought scientists and government officials together to discuss the implications of the new climatological insights. It was chaired from the beginning by the eminent Swedish climatologist Bert Bolin, who skilfully directed the discussions towards scientific consensus and political action. The IPCC and its excellent international reputation greatly helped to spur the debates of another body, the Intergovernmental Negotiating Committee (INC), set up to prepare a convention on climate protection.

A few weeks before the Earth Summit, the INC concluded its deliberations and negotiations on the Framework Convention on Climate Change (FCCC), so that it was available for signature in Rio. One hundred fifty-four states signed the Convention during the Earth Summit. Many heads of state or of government, including President George Bush, Prime Minister John Major and Chancellor Helmut Kohl, prided themselves on personally signing the Convention in the presence of the international press. It was perhaps the moment of highest media attention to date to the problem of global warming.

The Framework Convention is remarkably strong in its Article 2. It asks signatory states to stabilise greenhouse gas 'concentrations at levels preventing a dangerous human interaction with the climate'. This is quite a statement. If present trends continue we are certain to reach CO_2 concentrations very dangerously interfering with the global climate.

Still, much room is left for interpretation of what Article 2 means in practical terms. Even the signatory states seem unwilling so far to make major commitments with regard to agreed-upon goals for a reduction of greenhouse gases. Two Conferences of the Parties (CoPs) to the FCCC were held, in 1995 and 1996, respectively. The first one, in Berlin, produced the so-called Berlin Mandate to formulate a protocol which should be ready for adoption at the third CoP in Kyoto, Japan, in December 1997. While the Berlin CoP was seen by observers as a very frustrating meeting because the oil-exporting countries together with the US and Russia blocked nearly all progress, the second CoP held in Geneva was surprisingly successful in concretising the mandate for Kyoto, chiefly due to a turnaround in the attitude of the US delegation.

The strongest interest in climate protection understandably comes from the AOSIS group of countries, which fear sea-level rise and the further increase of extreme weather conditions – both likely consequences of global warming. Their draft protocol was, however, barely discussed at the Berlin Conference.

In the absence of diplomatic successes, let us try to understand what would be needed if the world were to take Article 2 of the Convention seriously. The international climatological consensus (see, *e.g.* IPCC, 1990, 1996) seems to suggest that to stabilise our climate would require reductions of greenhouse gas emissions by some 60 per cent worldwide. Imagine what that means for the industrialised countries! If we assume, as the World Energy Council (1993) does, that energy demand in developing countries will more than double, and that most of this demand is going to be met by fossil fuels, the industrial countries would have to reduce their CO_2 emissions by something near 80 per cent. Perhaps some 50–80 years may be available for reaching this goal. A graphic simplification of this situation was shown in Figure 1, p xxviii.

The challenge seems enormous. How can industrialised countries possibly reach a reduction of greenhouse gas emissions by a factor of five? Knowing about the difficulties, not to say impossibility, many authors (*e.g.* Nordhaus, 1993) recommend putting a much higher priority on adaptive rather than preventive measures, which they consider costly and futile. They fear that negligence with regard to adaptive strategies could lead to extremely high costs in the event of climatic change accompanied by rising sea levels, droughts or extreme storms. Let us not deride this attitude. It is one of the serious options available to industrialised countries.

There may be other options. One could be a dramatic cultural change in the North towards modesty and austerity. But historic experience does not seem to make this a very plausible option. Another possibility, discussed in more detail in Chapter 10, p 251, is a switch from fossil fuels to nuclear and renewable sources of energy. The outcome of that discussion is that fuel switches driven by necessity mean costs, in some cases very high costs, and have a rather limited scope.

We feel that the most attractive option by far is the efficiency revolution as presented in Parts I and II of this book. The Factor Four option, as emphasised earlier, is all the more attractive in that much of it is available at negative cost, *i.e.* profitably.

This presents a totally new situation to the otherwise tedious and frustrating climate diplomacy. Environmental diplomacy from its inception has been loaded with the assumption, not denied by any party, that costs have to be incurred for any ecological progress. If, in the case of efficiency, it is negative costs, or benefits, that accrue to those actively protecting the climate, then negotiations should be immensely facilitated.

The outcome of the Berlin meeting, however, seems to show that hardly anybody has so far understood the explosive opportunities lying in this new insight. It is to be hoped that the vague Berlin mandate for further study of a binding protocol will be used specifically for elaborating efficiency-oriented win-win strategies.

Non-CO_2 Greenhouse Gases

Even the Factor Four transition in energy productivity will not be sufficient to halt global warming. Greenhouse gases other than CO_2 contribute some 50 per cent to the greenhouse effect, including CFCs, methane (CH_4), water vapour (at high altitudes), N_2O, and ozone (O_3).

An essential contribution comes from the phase-out of CFCs to protect the stratospheric ozone layer. This is one of the most encouraging success stories of environmental diplomacy (Benedick, 1991).

Methane stems from rice paddies, cattle digestion and biomass degradation (composting). N_2O emissions also result from biomass degradation and from most combustion processes. N_2O remains in the atmosphere for a very long time, more than 150 years, while methane has a half-life in the atmosphere of 'only' 14 years. Changes in agriculture would be the most important clue to reducing methane and N_2O emissions. But as a reduction in food production is out of the question in a period of growing population, here, too, we may need a revolutionary increase in efficiency allowing us to produce more food with less methane.

8.4 SPECIES EXTINCTION AND THE BIODIVERSITY CONVENTION

The other major convention signed at the Earth Summit deals with biodiversity. What are the issues there and what can be done? Harvard biologist Edward Wilson has said the destruction of biodiversity is the sin for which future generations will forgive us least. And he has compared the life science community, faced with the present ecological destruction, with art lovers seeing the Louvre and other museums going up in flames but unable to stop the fire.

Biodiversity also serves as the ultimate ecological resilience against unpredictable climatic change or other alterations of the biosphere. It is foolish and highly irresponsible to sacrifice biological diversity to short-term economic advantage.

It was the Global 2000 Report to President Jimmy Carter (Barney, 1980) which brought home to America and the world that biodiversity losses had become starkly dramatic. Figure 28, a graph from Gerald Barney's institute, shows the dynamics of species extinction.

The reasons for this massive acceleration of species extinction are complex. Perhaps the single most important factor may be the debt crisis which has induced developing countries to sell cash crops, ores, timber and hydroelectricity to foreign countries and to foreign companies at unsustainable rates. Many of these export-oriented developments were carried out at the expense of virgin forests and other natural treasures and thus contributed greatly to the destruction of natural habitats. As nearly all developing countries expanded their resource exploitation and their exports at the same time, world market commodity prices fell drastically. As a result, the indebted countries had to sell even more of nature's treasures in order to maintain the servicing of their debts. Figure 29 shows the twin developments of Third World debts and falling commodity prices from the mid-1970s to the early 1990s. (A relief may be seen in the more recent uptick of commodity prices.)

The Biological Diversity Convention adopted during the 1992 Earth Summit of Rio de Janeiro, was also prepared by an intergovernmental negotiating committee. It was very much a convention pushed by the North, while the South was extremely nervous about maintaining its national sovereignty over its biological resources.

Since the 1972 Stockholm UN Conference on the Human Environment, developing countries had been highly suspicious about Northern ambitions to interfere in their ecological affairs.

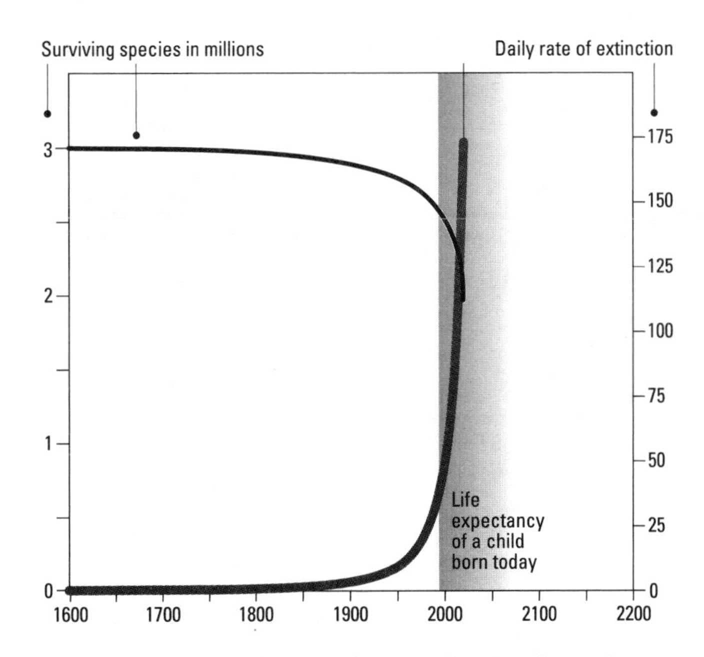

Figure 28 The extinction of species has accelerated to an absolutely alarming extent. Some 20–50 species may be lost daily! (Courtesy Institute for 21st Century Studies)

Principle 21 of the Stockholm Declaration, the precursor to the Rio Declaration, re-affirmed that 'States... have the sovereign right to exploit their own resources pursuant to their own environmental policies.' In Rio, the Third World went a step further and made it (in Principle 2) 'pursuant to their own environmental *and developmental* policies' (emphasis added). Their suspicion increased when the US, the UK, Australia and others involved in the negotiations about the Biodiversity Convention insisted on maintaining the property rights of germ plasm extracted from developing countries but biotechnologically processed in Northern laboratories. To the developing countries, it seemed that the entire interest of the North in biodiversity protection was related to exploitable genetic resources. The South nevertheless yielded in almost every conceivable respect. In particular, the South did not even mention the debt crisis, knowing that here was the most intractable cause of the dramatic biodiversity losses. True enough – the convention could do nothing to stop the debt crisis.

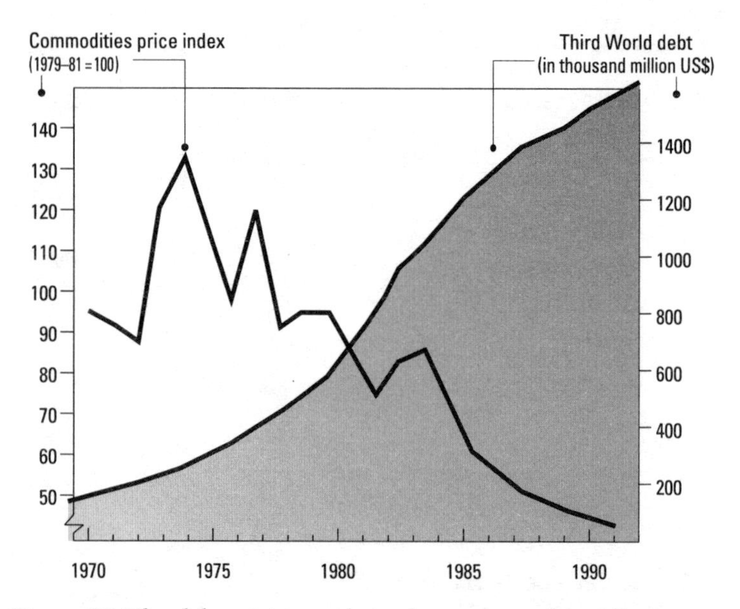

Figure 29 The debt crisis stands in the context of rapidly increasing commodity sales by developing countries and the ensuing collapse of commodity prices, and may be the single most important cause of biodiversity destruction. (Adapted and updated from Worldwatch Institute's 1990 State of the World Report, *p 144)*

At least some incentives were built into the convention for the developing countries. It was established that the benefits of biotechnological use of genetic resources should be shared on a fair and equitable basis (Article 19) and the North should provide new and additional resources to enable the South to meet the cost of implementing the agreed-upon obligations (Article 20). These relate partly to the preference given to *in situ* conservation, meaning protected natural areas (Article 8) over *ex situ* conservation, meaning botanical or zoological gardens or gene banks (Article 9). This is very prudent, since a hectare of virgin forest contains immeasurably more biodiversity than all that can be cultivated at Kew Gardens plus the accumulated breeding results of all biotech laboratories of the world.

The first CoP was held in Nassau, Bahamas, in November 1994, but produced no major breakthrough towards making the language more concrete. At the second and third CoPs in Jakarta (1995) and Buenos Aires (1996), at least the need for a biosafety protocol was agreed to. The delegations from the developing world have begun to see the nexus between habitat protection and the absolute need, thus far largely ignored by the North, for regulating biotechnology and the use of genetic material in agricultural and pharmaceutical applications (Third World Network, 1995).

The reader may nevertheless ask what the relation could be between Factor Four and the preservation of biodiversity. In fact, the efficiency revolution is bound to help conservationists a great deal. Reducing the demand for electricity would come as a great relief to environmentalists in both the South and the North. Reducing the massive extraction of material resources such as minerals and wood fibre could even more become the anchor for strategies to protect the remaining wildlife habitats.

What is more, an efficiency revolution can also be conceived for the food and land-use sectors. Quadrupling land productivity is not theoretically more difficult to imagine than quadrupling energy productivity. There is theoretically no need for shifting cultivation in virgin forests (a method that was sustainable for a very thin population density, but which has become one of the most devastating practices in recent years, partly because of the massive population increases and partly because it changed its face to become cash-crop oriented with little regard to the special properties of the soil).

We chose not to include a chapter on land productivity because it is in no way as simple as resource productivity. In many cases, the only sustainable solution is extensive, 'unproductive' land use. Nevertheless, many of the examples of Part I of this book mention the potential for better land use to the benefit of biodiversity.

8.5 More Unsolved Ecological Problems

Many more ecological problems can be cited as causing concern. We have no intention of trying to name them all. The most serious will be dealt with in the next chapter – the swelling flow of materials that is disrupting the surface of the earth. Of the other problems we will mention only two more, eutrophication and overfishing.

In the 1960s and 1970s, eutrophication was identified as one of the biggest problems of water pollution. Phosphates and nitrates had transformed many lakes and slow rivers into a state of overnutrition and ensuing algal growth. Fish and other animals that needed high levels of oxygen disappeared, and in many cases the ecosystem collapsed, leaving extremely unpleasant odours (and toxicity) behind. Municipal wastewater cleaning and phosphate-free detergents helped greatly in the more advanced countries to improve the situation, and eutrophication more or less disappeared from the environmental policy agenda.

That was premature. Eutrophication is still looming large, if in different contexts. One is marine eutrophication. Notably the Black and Baltic Seas are severely in danger of being fatally eutrophied. As a matter of fact, the deeper layers of the Black Sea have been dead since prehistoric times, owing to the steady influx of nutrients from the Danube, Dniestr, Dniepr and Don rivers and the lack of vertical currents. The dead and sulphuric deep-sea environment make for the black appearance from which the name was taken. In postwar times, however, the massive additional flows of nutrients from East European agriculture and human settlements made the boundary rise between the dead water masses and the living surface layer. The hydrogen sulphide stench now regularly comes to the surface and has begun to threaten riparian residents and tourism. Similar developments are feared in the Baltic Sea.

Another form of eutrophication comes through the air. In Central Europe, nitrous oxides from car exhaust, ammonia from agriculture and a variety of other nitrous emissions contribute to an unprecedented nitrate load in the air. This leads to nitrate depositions often of 50–100 kg of nitrate per hectare per year, which is more than is usually applied by farmers when fertilising their fields and meadows. Nutrient-rich meadows, on the other hand, are deplorably poor in biodiversity. Fast-growing plants like dandelion, coltsfoot and common grasses dominate the ecosystem and overgrow all the delicate nutritional specialists. Also, groundwater is increasingly affected, making the new wave of eutrophication a major danger to drinking water reserves.

We report on eutrophication not only because it is an important problem, but also to warn against simplistic methods of efficiency technology. Farming efficiency and land productivity gains contribute to eutrophication. The answer will lie in modernised high-efficiency organic farming (see Chapter 1, p 50 and Chapter 2, pp 97–101).

Overfishing is yet another unsolved problem. The *1995 State of the World Report* of the Worldwatch Institute (Brown *et al.*, 1995) shows clearly that world fisheries have surpassed the maximum sustainable yield. Prices for fish have been going up for years and incite the fleets to ever greater efforts at increasing or maintaining the catch. But in vain. The fish stocks need time for recovery. Massive unemployment in the fishing industries is being

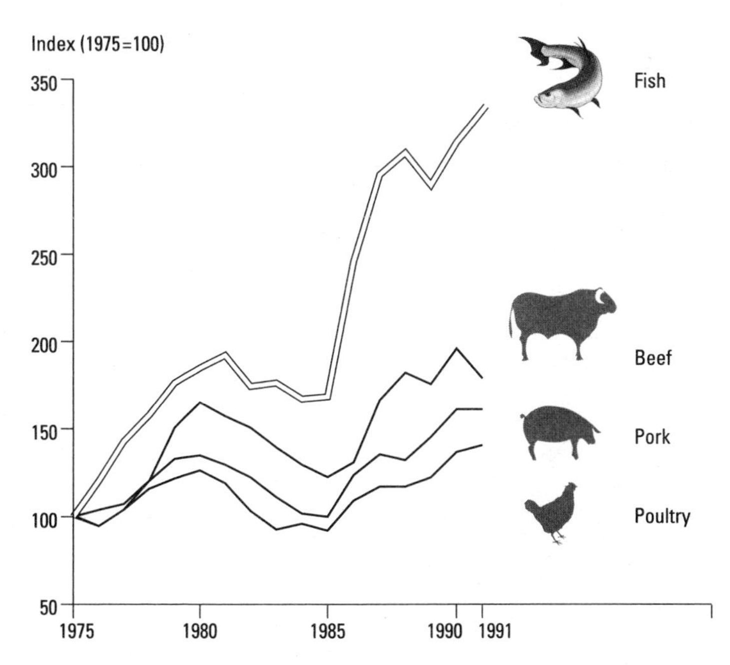

Figure 30 Stagnating world fish catches and rising demand lead to exploding fish prices. Fish has been a staple food for the poor for centuries, yet is now becoming as precious as the highest-quality meat (The vertical axis shows real prices.) (Brown-Weiss, 1989)

forecast as a combined result of stagnating or shrinking catches and still ongoing mechanisation. Figure 30 shows the stagnation and price explosion during the past ten years.

Where are the solutions? Not in further mechanisation of ocean fisheries; rather in producing fish in combination with local agriculture. High-yield integrated systems exist. Fish can feed on algae that live on organic residues from farms. A promising initiative was taken in February 1996 by the World Wildlife Fund and Unilever to create a brand name for sustainable fishery, to be controlled by a Marine Stewardship Council. This would allow customers to help support sustainable fish stock management by boycotting suppliers not subscribing to the council's rules.

We conclude this chapter with a warning about surprises. Some problems are invisible in their early stages of development. William Stigliani, now with Iowa State University, formerly with IIASA (the International Institute for Applied Systems Analysis in Austria), has

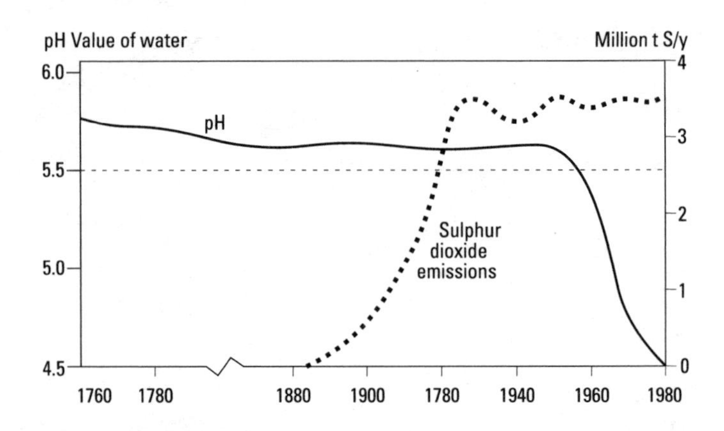

Figure 31 One instance of an ecological time bomb. It took 50 years of acid rain for Big Moose Lake to be pushed beyond its limits (courtesy William Stigliani). But no harm was evident at first.

shown the temporal dynamics of the acidification of Big Moose Lake in upstate New York. His findings are depicted in Figure 31. For more than 50 years, increasing quantities of sulphur in the form of sulphuric or sulphurous acids rained down on the region, until finally the buffering capacity of the soil in the catchment area and of the lake itself was exhausted. The lake was eventually pushed past the point of no return, since when it has been virtually dead.

Chapter 9
Avalanches of Matter: The Forgotten Agenda

It's a moving story. We are moving the earth. We humans move more earth than do volcanoes and the weather combined (Schmidt-Bleek, 1994, p 37). By moving things round we do a lot of damage. We overstress the earth's capacity safely to absorb the thousands of millions of tonnes that we return as waste or as overburden. The avalanches of matter may turn out to be the greatest threat to the global environment. 'In the past, environmental protection has targeted nanograms. Now it's high time for us to look at the mega-tonnes,' says Friedrich Schmidt-Bleek, a pioneer, along with Bob Ayres and John Young, in making materials – toxic or not – a major issue of environmental policy.

The Club of Rome was also among the pioneers with its vision of an age *Beyond the Age of Waste* (Gabor and Colombo, 1976). Regretfully, this important Report to the Club of Rome aroused very little interest. Perhaps the authors were just unlucky in addressing a public that was too much preoccupied with the energy crisis and the collapse of the Iranian Shah's regime to pay attention to the visionaries of a lean economy.

What kind of problems do we cause by moving and transforming the megatonnes? Well, what we have noticed first in the industrialised countries is the waste problem. Landfills have been the cause of endless quarrels in and around American cities. For a while, incinerators were welcomed as a solution. But soon they themselves became the targets of fierce environmental opposition. Recycling and remanufacturing were the logical next steps. But progress remained slow in this regard, and the avalanches of waste kept growing. The nice side of this, from an economic growth point of view, is that waste management has become a flourishing multi-thousand-million-dollar business with perhaps in the order of a million jobs in the US alone. But ecologically this offers little relief. Waste management doesn't work on the earlier part of the material streams which environmental policy has had a tendency to forget, and which are not at all innocent from an ecological point of view.

9.1 THE WASTE PROBLEM IS ONLY THE TAIL END

Whenever a problem has a high profile on the political agenda, the temptation is to throw big money at it. This is what happened with the waste issue. But simply throwing money at a problem does not normally lead to elegant solutions. Not surprisingly, present-day waste management is less than elegant.

Waste management in OECD countries is founded on a sizeable amount of legislation. By making waste treatment and waste disposal ever costlier (and ever more detrimental to a company's public image), a commercial situation was created where 'waste reduction always pays' (WRAP), to quote a popular slogan (and programme) of the Dow Chemical Company. Industry in the US, Japan and Europe has indeed managed to reduce waste outputs greatly. This was in most cases achieved by internal recycling of materials, *i.e.* decreasing the material intensity of production.

One of the problems that remained very difficult to treat, however, was packaging waste. Packaging has become the symbol of the modern consumer society. Supermarkets believe they need handy, hygienic, attractive packaging for virtually all goods they sell. Industry is not in a position to change this view. So recycling the packaging waste appears to be the only way of increasing the productivity of the materials. But then, much of the packaging material doesn't lend itself to easy recycling. Compound materials in frequent use contain various mixtures of plastic, metal, paper and wood. How can this be handled?

In 1991, Germany moved into the lead with a bold ordinance on packaging waste that obliged manufacturers, domestic as well as those exporting goods to the country, to pay a fee for the collection and recycling cost of their packaging. Payment of the fee earned products the Green Dot. Differentiated fees (since late 1993) ensured that an incentive was created for lightweight, simple and easily recyclable packaging materials. More than a million tonnes of municipal waste was avoided annually, according to the DSD, the private monopoly contractor firm handling the system. In the end, of course, consumers paid much of the bill and became angry at the whole system. The DSD (Duales System Deutschland) became almost a symbol for ecological nonsolutions. (Chapter 2, p 104, presents more detail on the shortcomings of the Green Dot system, and suggests a high-tech solution for the recycling of plastics used for blister packaging.) At any rate, the heated controversies about DSD brought it home to the German public that end-of-life-cycle solutions and incentive systems are not very attractive in the waste context.

Around the time of the strongest public criticism of the Green Dot system, the German parliament, the Bundestag, had finally reached agreement, after years of controversy, on the establishment of a Special Committee of Inquiry (Enquête-Kommission) on chemicals. The environmental camp had a long-standing intention to analyse and castigate chlorine production and other activities seen as the origin of toxic waste and pollution. When the committee finally started its work, in 1992, the lawmakers were already a bit bored with yet another inquiry into preventive control of nanograms of toxic chemicals, and began to realise the need for a comprehensive approach to life-cycle material flows. It was becoming apparent to all policymakers that the unpleasant waste problems were only the tail end of a long and complex story, which had not thus far been grasped by environmental policy.

In this situation, the Bundestag Committee took a major interest in Friedrich Schmidt-Bleek's ideas of global material flows and 'dematerialisation'. After two years of work they produced a major volume of findings and policy proposals (Enquête-Kommission, 1994) which can be seen as a political start for the idea of material efficiency.

Looking at the Megatonnes

Perhaps society has been looking at material flows from the wrong end altogether. At the front end, we allowed technological progress to make prospecting, mining, processing and shipping of material resources ever more efficient and ever cheaper. As a result, the sheer amount of materials arriving at the gates to our civilisation kept growing. Small wonder that the waste problem was building up. To find a solution that does more than scratch the surface we have to do more than waste management. We have to address, control and decrease the material streams that are flowing in.

Earth movements and human-induced changes of water flows are the biggest among the material flows that need to be addressed in the first place. Wilhelm Ripl (1994) has shown that earth movements and drainage have begun seriously to erode the cation (positive-ion) concentrations in soils and watersheds. By definition this leads to acidification. Ripl holds that the causes for soil and water acidification lie more with earth movements than with acid rain. Figure 32, a qualitative drawing by Ripl based on Scandinavian studies of soil acidification, indicates an absolutely dramatic loss of cations in recent years. Nearly all the reserves built up in the Ice Age have been used up in just a few decades.

Earth movements also have another effect on acidification. Peter Neumann-Mahlkau (1993) reports that more sulphur dioxide is created by the mobilisation of atomic sulphur through soil movements than by all industrial combustion processes taken together.

Tropical forests may suffer more from mining (including access roads) than from logging for lumber. For the most part, mining, processing and shipping of material resources is the source of both local environmental degradation and greenhouse gas emissions that relate to transport and energy (Young, 1992).

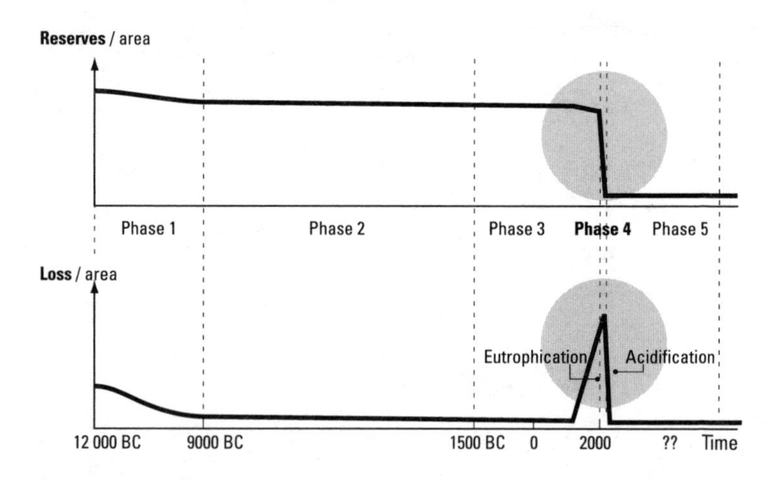

Figure 32 A large stock of cations was built up in Scandinavian soils and was maintained over the postglacial millennia. A hundred years of earth movements and drainage drained so much cation stock off the soil that lasting acidification followed, with a severe danger of desertification. (Qualitative drawing by Ripl, 1994)

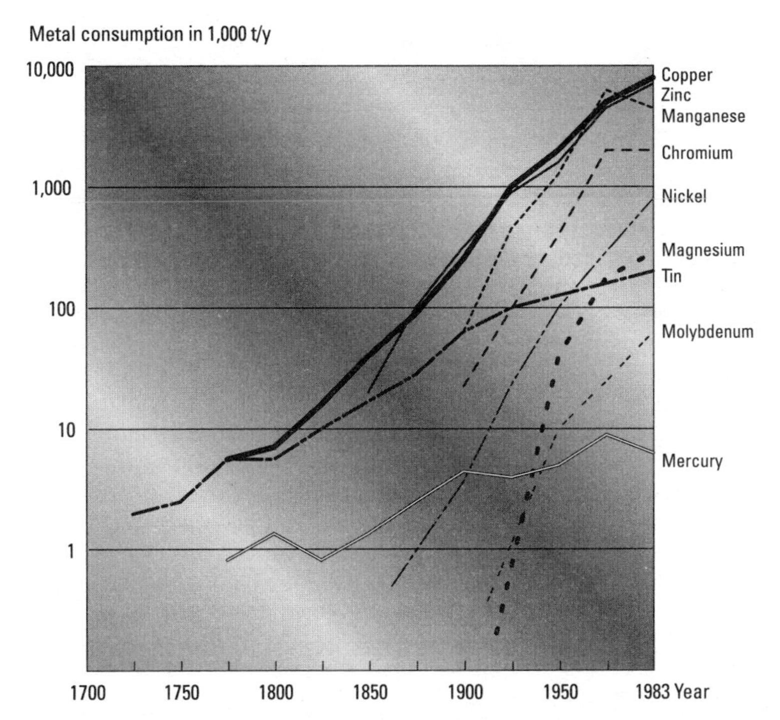

Figure 33 Exponential growth in the use of metals by our civilisation has been the pattern for decades. (After Ayres and Ayres, 1996, p 7)

And the avalanches of matter are growing. Robert Ayres (1996, p 7) has shown that exponential growth was involved in the use of nearly all metals (Figure 33). This avalanche of matter coming into use in our civilisation is but the tip of the iceberg. Each tonne of a metal involves many tonnes of ores that have to be mined, purified and processed. Gold mines, for example, are known for causing enormous earth movements and large-scale water pollution with heavy metals other than gold (not to mention the cyanide often used to extract the gold). What is left, mostly near the mine, is labelled 'overburden' (rock that had to be stripped away to get at the underlying ore-bearing formation), or 'tailings' (ground-up rock left over from the chemical extraction of the ore-bearing mineral) – both terms indicating that we consider these huge masses of rock to be useless. Often such leftovers are poisonous to plants and animals. Greening the tailings is not a trivial matter and will hardly ever restore the virgin aquifers, habitats and landscapes affected.

9.2 THE GOLD RING ON YOUR FINGER WEIGHS THREE TONNES

Material efficiency has to be defined somehow. Schmidt-Bleek (1994) has made a bold attempt by establishing the material intensity per service (MIPS) concept. For all types of goods and services, he and his team estimate and calculate the material flows in tonnes on a cradle-to-grave basis. They do it for the gold ring on your finger, for the daily newspaper, for orange juice and for autos. As goods tend to represent the availability of services, it is ultimately services that interest the user of goods. It is the mile driven in a passenger car or the convenience of sitting in a comfortable chair or the 'service' of showing marital status by a gold ring that would ultimately be used as the denominator in the MIPS calculation. (The definition of a service actually turns out to be the most difficult part for Schmidt-Bleek's MIPS calculation team!) Longevity of goods has a positive influence on MIPS. By definition, it decreases the MIPS of the respective services. Chapter 2 has presented a number of examples of reducing MIPS by a factor of four or more and thereby of increasing the material productivity correspondingly.

Schmidt-Bleek's team goes as far as using MIPS as a generalised yardstick for the ecological impacts of services. Of course, it is acknowledged that other and more refined yardsticks are available. There is toxicity, there is land use, there are greenhouse gas emissions. But in some way or other, all relate to the intensity of material turnover. And it is extremely valuable to have something simple for the 'quick-and-dirty' assessment of ecological impacts.

A kilogram of metal obtained from mines often requires tonnes of ores to be processed. Schmidt-Bleek (1994) speaks of the '*ecological rucksack*' carried by the metal. Figure 34, modified after Schmidt-Bleek, shows the relations between the metals obtained and their ecological rucksacks.

In the case of gold and platinum the relation is 1:350,000. Imagine the weight of the ecological rucksack of the gold ring on your finger. It would be 3 tonnes for a ring weighing a mere 10 g.

Metals were just the beginning of Schmidt-Bleek's rucksack story. Energy, too, carries a rucksack. Those 3 thousand million tonnes of coal that we burn each year carry a rucksack of tailings and water weighing easily 15 thousand million tonnes – not to mention the 10 thousand million tonnes of CO_2 that are released in the burning process. The relations are worse for lignite. Here the rucksack is ten times heavier than the lignite obtained.

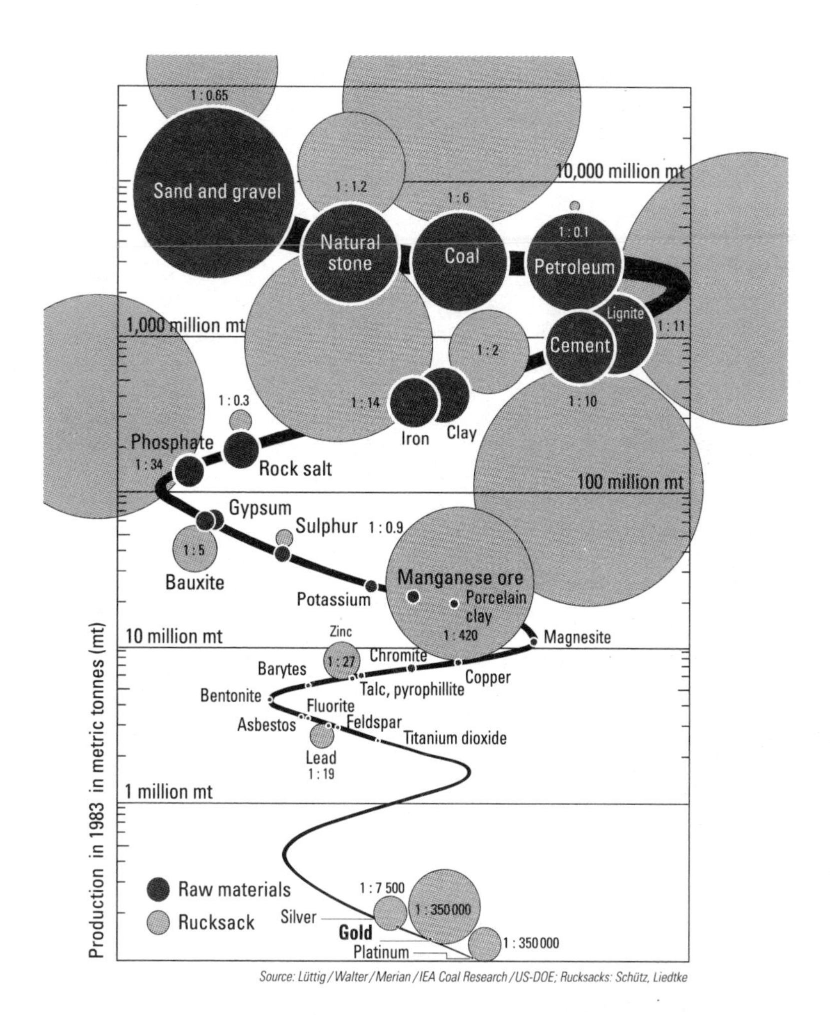

Figure 34 A kilogram of metals may require tonnes of materials to be moved and processed. Schmidt-Bleek (1994) calls this the 'ecological rucksack'. Some examples are given of ecological rucksacks typically carried by different metals. Gold is among the worst.

Every good and service that we enjoy has an ecological rucksack to carry. The catalytic converter in cars, once heralded as the saviour of German forests, weighs less than 9 kg but carries a rucksack of more than 2.5 tonnes, chiefly owing to the platinum that is used. (Obviously that rucksack shrinks by carefully recycling each milligram of the platinum, and it is worth doing so.) Orange juice isn't innocent either. Depending on the country of origin, a litre of orange juice has caused soil and water movements of more than 100 kg. Your daily newspaper may weigh a pound, but its ecological rucksack is likely to be 10 kg. Manufacturing a car typically involves 15 tonnes of solid waste, not counting the water that is used and polluted in the process.

But a Factor Four improvement in material productivity would make ecological rucksacks lighter by that factor. The examples in Chapter 2 represent much relief from the unwanted burden. The trouble is that a Factor Four may just not be enough for ecological sustainability. Schmidt-Bleek thinks it should be a factor of ten at least.

9.3 THE FACTOR TEN CLUB

Looking at the total impact of human interference with the biosphere, Schmidt-Bleek, like others (*e.g.* Rees and Wackernagel, 1994; Weterings and Opschoor, 1992), comes to the conclusion that material turnovers should be reduced by at least some 50 per cent on a worldwide basis. Since per capita consumption is something like 5 times higher in OECD countries than in developing countries, and further increases in world population are unavoidable, he says that sustainable levels of material flows will not be reached unless and until the material intensity of the OECD countries is reduced by a factor of ten. This is then the material target for sustainability.

Based on these considerations, Schmidt-Bleek has taken the initiative of founding the 'Factor Ten Club' of prominent environmentalists subscribing to that goal (see Plate 13). The principles of the club were laid down in the Carnoules Declaration of October 1994. Published in five languages including Japanese, the declaration calls for an efficiency revolution desubsidising resource use, and a new perception and definition of welfare.

Among the club members are Jacqueline Aloisi de Larderel, director of UNEP's Industry and Environment Programme; Herman Daly, formerly at the World Bank; Ashok Khosla, president of Development Alternatives, India; Jim MacNeill, the former executive director of the Brundtland Commission; Hugh Faulkner, former executive director of the Business Council for Sustainable Development; Richard Sandbrook, executive director of the London-based International Institute for Environment and Development; Wouter van Dieren, who has produced for the Club of Rome the new Report *Taking the Earth into Account* (van Dieren, 1995); and Ernst von Weizsäcker, coauthor of this book.

The Factor Ten Club initiative has immediately impressed some technology officials at the European Commission. Peter Johnston and Robert Pestel (son of Eduard Pestel, the late cofounder of the Club of Rome), at the Directorate on Advanced Communications Technologies and Services (ACTS), are working on a memorandum of understanding for 'An Information Society for Sustainable Development'. For this, they see the factor ten postulate (perhaps with some compromise formulations) as a highly inspiring challenge, and they want major European enterprises to sign and support the memorandum.

As was said in the introduction to Chapter 2, an improvement by a factor of ten may be easier to achieve in materials than in energy. This has to do with the physical stability of most materials. As long as their dissipation is prevented, they can be used again and again. One of the biggest enemies of material efficiency, therefore, is waste incineration, among the most dissipating waste disposal technologies. Still, many short-sighted people herald it as the most convenient 'solution' to the waste problem. It is a solution to the problem of toxic organic substances in industrial and municipal wastes – where the creation of those substances cannot be avoided in the first place by redesigning the product or process. Yet, under the factor ten obligation, waste incineration is rather an antisolution.

Chapter 10
Unsatisfactory Part-Solutions

10.1 Costly Pollution Control: Steering from the Wrong End

Rachel Carson was a heroine. Singlehandedly, she alerted the public to one of the biggest scandals of this century, the chemical pollution of the world. Her book *Silent Spring* became a milestone of environmental politics. Starting with pesticides, Carson's main target, pollution control became a synonym for environmental protection. Reacting to the new public demand, the US Environmental Protection Agency was established and the Clean Air Act, the Clean Water Act and more pollution control legislation were adopted at federal and state levels.

Similar events happened in Japan, where the deadly Itai-Itai and Minamata diseases, caused by heavy metals in drinking water, aroused the public. Widespread lung diseases caused by heavy air pollution made environmental protection a major public issue in all affected areas. As in America, pollution control was the obvious remedy. Western Europe also had its years of pollution alarms and developed controls by means of new legislation and enforcement of old legislation.

The environmental protection movement got its political strength from local initiatives, which sprang up all over the US and later in other democratic countries. Activists took the fate of their neighbourhoods into their own hands and thereby became a symbol of a rejuvenated democracy. With their sense of rebellion, these activists wanted to distinguish themselves from old-fashioned conservationism, although later the two streams merged into the broader environmental movement.

Millions of lives have been saved or made much more pleasant in the prosperous countries as a consequence of pollution control. There can be no question of abandoning the benefits of this successful new branch of civilisation. But there are limits to pollution control.

Environmental protection was an uphill struggle from the beginning and will remain so for many years to come. The simple reason is that environmental protection, like nature conservation, was seen as an economic sacrifice. Adding filters or more sophisticated devices to the existing production process invariably meant additional costs to manufacturers. They were forced by legislation and by litigation to incur these costs. But they could always argue that there were narrow limits to their ability to pay. They suggested they would be forced to emigrate if the costs and risks rose too high.

To be sure, economists began to realise that a healthy environment was a social asset worth paying a price for. And environmental protection businesses emerged that created more than 2 million new jobs for these new tasks in the US alone (Renner, 1991). Some traditional companies also began to realise that they fared better economically by pollution *prevention*. However, they did so under certain conditions related to avoiding or reducing litigation costs, to corporate image and identity, to modernising the capital stock and, in some cases, to financial savings in raw materials. However, these conditions depend on legislation, enforcement, the legal 'culture' and on commodity prices. Such conditions vary from country to country. Companies will always be forced by competitive pressures to compare costs and benefits for different places in the world and to opt for the most favourable place.

In other words, despite all the well-known reasons in favour of pollution control and pollution prevention, pollution control often means, and pollution prevention may mean, added costs. As the case of Dow Chemical shows, the costs are not always distributed very sensibly either. Their estimates for one new rule are that 80 per cent of the company effort will go for documentation and 20 per cent for monitoring and maintenance of equipment. The cost increases are undeniable. Thus, we can understand why less-developed countries consistently say they cannot 'afford' pollution control. This is tragic for the people in the population centres of the Third World, where vast numbers of people have to make do with grossly inadequate water and air.

The tragedy goes far deeper and must concern us in the rich countries. If countries required an OECD level of affluence to afford pollution control, then the whole game would be lost. Imagine what would happen to the world if all 5.8 thousand million people were rich enough to 'afford' pollution control. It would be the ecological ruin of the world, since OECD affluence is characterised by levels of resource consumption easily 5 times, often 20 times, higher than in developing countries (remember Figure 22). And even today's global consumption rates are clearly unsustainable.

And yet, despite the blatant unsustainability of the pollution control paradigm, rich countries go on with it and even proceed towards what may be considered a physically impossible goal: 'zero'-emission factories and machines. If classical pollution control is unavailable to 80 per cent of the world, zero emissions are unattainable for 98 per cent. The best justification for the zero-emissions approach could be technical breakthroughs elicited by happenstance in the effort to meet extreme pollution control standards. In the case of the Californian car exhaust stipulations, for example, they may actually encourage the accelerated introduction of the 'hypercar' (Chapters 1, p 4 and 2, p 71) which happens to be also very much cleaner than its conventional gas-guzzling rivals.

PROFITABLE ENVIRONMENTAL POLICIES

This book intends to show the way out of the pollution control dilemma. The Factor Four message means a departure not from pollution control as such, but from the exaggerated emphasis on it and from the philosophy of costly environmental policy. Our goal is to make environmental policy a fundamentally profitable undertaking. For this, we are shifting our attention back to resource questions.

Why 'back'? Well, remember the 1970s! When *The Limits to Growth* report to the Club of Rome was published (Meadows *et al.*, 1972), people around the world up to the highest echelons of power were impressed. This report laid strong emphasis on the exhaustion of scarce resources if current trends were to continue. A year later, the oil price shock seemed brutally to confirm the scarcity fears. As a result, resource scarcity and rising resource prices became a matter of top public concern.

But there was something many people overlooked. In response to high commodity prices and widespread fears of scarcity, resource exploration and exploitation began to boom in an unprecedented manner. New oil and gas fields were discovered almost every month. For 15 years in a row, far more oil and gas reserves were discovered annually than were burned (Yergin, 1992). As we noted in Chapter 9, p 249, high-tech geological exploration made access to fossil fuels in the ground and to other mineral resources ever cheaper. These trends continue today and are even accelerating.

Not surprisingly, the bargaining position of oil producers eroded quickly, and by 1986, the world market price of oil and gas had dropped to levels lower than 1973 before the oil crisis. Similar developments occurred in other commodity markets. By the mid-1980s, resource scarcity had virtually disappeared as a public theme.

Just about that time, however, a new 'scarcity' was discovered (or rediscovered): the absorptive capacity of the biosphere for all the pollutants that were emitted no matter where (see Cairncross, 1991). The symbolic substance for that discovery was CO_2. The greenhouse effect (Chapter 8, p 222) seemed to force us to reduce CO_2 emissions drastically. Philosophically, the talk about the absorptive capacity brought us back to the early days of emission control and pollution control. Except for the fact that it seemed difficult to conceive of CO_2 filters, greenhouse gases were seen as pollutants to the atmosphere. And economists like William Nordhaus jumped on the new theme of pollution control using their standard repertoire of pollution control economics. Not surprisingly, they 'discovered' immense 'costs' to be incurred in the hypothetical battles against the greenhouse effect (Chapter 4).

Let us for the sake of the argument accept that we should be more concerned with emissions than with scarce resources. Even under this assumption, the question remains how best to tackle the emission problems. We believe that, for all practical purposes, it is far more promising to tackle the CO_2 problem and many other pollution problems by first optimising primary resource *use*.

The simple reason is that increasing resource productivity can be a highly profitable strategy, and 'picking up £20 notes from the street' has been our expression for much of this agenda. So, although we acknowledge (and share) the widespread view that we need not be overly concerned with resource *scarcity*, we are citing economic profitability and efficiency reasons for turning our attention back to controlling and diminishing the consumption of primary resources.

10.2 HIGH-TECH FANTASIES AND THE NEO-CORNUCOPIAN IRONY

Many people believe we don't need the efficiency revolution. Some may plainly deny that there is any ecological problem. Some admit there is a problem, but are convinced it will be solved the way problems have always been solved – by supply-side technical means.

To the problems of air and water pollution, technical answers have been found that are now called pollution control. If the new problem is CO_2 emissions, the answer should be CO_2 absorption, nuclear power or other carbon-free energy sources. If traffic congestion is the problem, better roads and electronic traffic guides will help.

The Japanese are particularly proud of their successes in using technology for beating problems – and beating their competitors at the same time. In the past, Japanese technology development served chiefly the goals of catching up in, and accelerating the process of, industrialisation. More recently, Japan has begun to launch strategic attacks on the global problems that were created or deepened by the industrialisation process.

The authors are certainly aware that some of the proposed high-tech options are not without major problems. Nuclear energy today represents only 5 per cent of the world energy pie (see Figure 35). Let us, for a quick assessment of the scope for nuclear power, very daringly assume that the total nuclear power-generating capacity can be tripled in 40 years. In doing so, let us generously ignore the financial, safety, military, terrorism, pollution and disposal problems that may be associated with this tripling of the nuclear power establishment. Let us also assume, as some Japanese futurists and the World Energy Council do, that energy demand will more or less double during that timespan. Then we would end up with a nuclear contribution to the world energy pie worth a mere 7.5 per cent! Would this be worth the price? It would certainly not prevent fossil-fuel burning from increasing further.

How about sinking CO_2 deep into the oceans? The yield factor will certainly be unsatisfactory. After all, the mechanical energy required for compressing and pumping would eat up some 50 per cent of the usable energy produced in the first place. Absorbing CO_2 from the exhaust gases of fossil power plants may be slightly more promising. Carbon dioxide introduced into big ponds can serve as a nutrient for algae growing in the ponds. But then the algae have to be harvested, dried and used either for another burning cycle (replac-

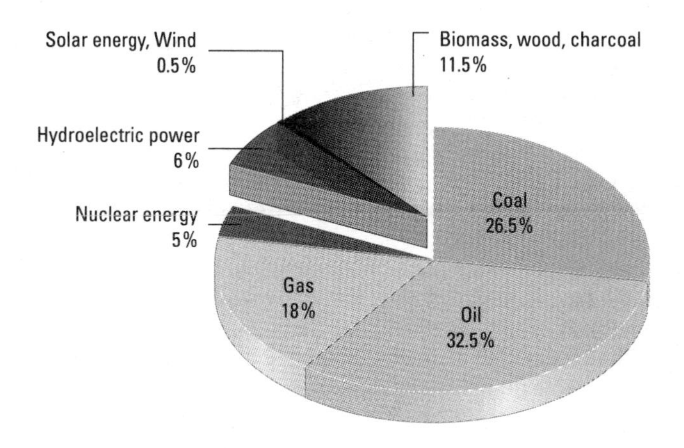

Figure 35 Five per cent of the world primary energy pie is nuclear – less than half the output from biofuels. Tripling nuclear energy, while at the same time allowing the pie to double, would increase the nuclear contribution to just about 7.5 per cent, while the fossil contributions kept growing in absolute terms.

ing some fossil fuels) or for final deposition – at a price. There's certainly no scope for a free lunch along these lines.

Carbon dioxide from the air is absorbed by green plants. A policy of massive afforestation on suitable sites makes a lot of sense (Read, 1994). However, not all sites are suitable. 'Greening the deserts' programmes may be ill-fated from the outset. Think only of the unavoidable salination of the soil if a steady flow of freshwater (containing certain quantities of salt) enters the system, but the only outflow is via (salt-free) evaporation.

More disquieting still are fusion and space solar energy. Fusion power is a very far cry from being safe and clean. It requires enormous quantities of radioactive hydrogen, causing unprecedented containment problems. Also, extremely high intensities of neutron flows are physically unavoidable with fusion. Neutrons travel through virtually any wall material until they join unpredictably in all kinds of atomic reactions, making the wall materials very unpleasantly radioactive. Fusion neutrons may even be used to make nuclear bomb materials. And nobody knows whether it will ever be commercially viable.

Space solar power isn't any better. Commercial feasibility may be still more questionable than in the fusion case: if we had the inexpensive solar cells required for the scheme to make sense, they'd probably deliver solar power at lower cost if simply installed on your roof. Moreover, solar power satellites could easily be the most dangerous source of energy in view of their suitability to military use and their susceptibility to terrorism that could either interrupt their highly concentrated power supplies or use them as weapons. Lots of science-fiction stories build on the potential of redirecting high-intensity power beams from satellites and blackmailing a trembling earth.

High-tech fantasies serve the purpose of attracting R&D money. The strongest advocates of fusion power are not found in the energy business but in the scientific and engineering communities. We don't begrudge them their money and their jobs, but why should they spend their time and taxpayers' money on such useless, costly and unprofitable things as fusion power? The spin-off in terms of new technologies and new scientific discoveries would probably be higher if the money went directly into applied basic research. And the economic and societal benefits would very likely be much higher if much of the money were redirected into the efficiency revolution!

Neo-cornucopian Irony

Much of the history of high-tech answers to existing problems is actually full of irony. Among the first answers to *The Limits to Growth* (as well as to Paul Ehrlich's *The Population Bomb* and Barry Commoner's *The Closing Circle*) were counterattacks from the scientific and engineering communities, most prominently represented by John Maddox, editor of *Nature*, arguably the most prestigious journal of the scientific world. He summarised his anger at these books in *The Doomsday Syndrome* (Maddox, 1972), written in a jolly and aggressive style. He suggested that, just as with earlier problems, the newly perceived problems related to population, development and environment would be readily solved in due course by the intelligent application of advanced technologies.

For such high-tech fantasies, Maddox (later knighted) and such followers as economist Wilfrid Beckerman were called 'cornucopians' by their critics, including one of the authors of this book (ABL). We felt that the fabulous new technologies, though no doubt very powerful, also had costs and side-effects. For this irreverence, the cornucopians dubbed us 'technological pessimists'.

Twenty-odd years later, some of the 'technological fixes' have proven more powerful than anyone had expected. However, the winners were *not* fast breeder reactors and solar power satellites, but rather microelectronics, miniaturisation and labour-saving production technologies. Also among the winners were many of the efficiency-related technologies which we describe in Part I of this book.

Looking at economic and technological realities with our own eyes, we feel that superwindows (Chapter 1, p 19), perennial polyculture (Chapter 2, p 97) or Belland material (2, p 104) now look superior to fast breeders, megafertiliser factories and palm-tree plantations cloned from tissue cultures. Gigantic water projects lost out to drip irrigation (2, p 80) and efficient household appliances (Chapter 1, p 29). Oil shale lost to mineral wool, heat exchangers and hypercars (1, p 4). In short, supply-side wizardry proved, in general, uncompetitive against resource-efficiency technologies. Some of the dark horses and unknown jockeys from obscure stables won, while the favourites broke down just out of the starting-gate. This has strewn the landscape with hundreds of thousands of millions of dollars' worth of decaying technocarcasses.

And the original roles are reversed too. It is the former 'technological pessimists' like us who are now the 'neo-cornucopians'! And it is the former 'cornucopians' who have turned into 'technological pessimists', asserting that our demand-side solutions don't exist, or aren't cost-effective, or won't sell or can't be relied on – so that all their costlier supply-side technologies, they assert, must still be bought as an 'insurance policy'.

It may not be possible to buy full sets of demand- *and* supply-side solutions, because they all compete for resources, and some actually preclude others. The supply industries need high demand to pay for their huge supply investments. Successful purchase of both increased supply and increased efficiency yields the worst of both worlds – the costs of the increased supplies without the revenues to pay for them. This scenario has actually occurred several times in the US since 1973, with the most unpleasant consequences for the oil, nuclear and utility industries: they had to write off hundreds of thousands of millions of dollars' worth of assets meant to produce costly energy that nobody wanted.

A deeper lesson, though, is that nobody is very good at forecasting what technologies people will want, invent, sell or buy. It is commonplace to observe that predictions tend to overstate the short-term effect and to understate the long-term effect of innovation. It is less often observed that major classes of innovation (hypercars are among the latest examples) were not expected at all, and resulted from the fortuitous and artful fusion of technologies separately developed for quite other reasons.

The case-studies reported in this book therefore use only technologies already developed, at least to the point of technical proof, and usually already on the market. But others not yet thought of – a category for which no examples can yet be given – are likely to make the task of resource productivity even simpler. It is now incumbent on the neocornucopians to avoid their predecessors' hubris while exercising their expansive vision.

10.3 Ecological Audits – Costly but Possibly Enlightening

Faithful information about a company's ecological performance is the most important prerequisite for green management. The beginning was made in the US in the 1970s. After the Love Canal disaster and other accidents, both the public and the insurance companies began to insist on a systematic analysis of potential dangers inherent in manufacturing processes. Forced by law, industrial firms began systematically to assess their accident risks, inform the relevant authorities and allow inspections from independent experts.

Very similar developments took place in Europe after the 1977 Seveso accident. The Council Directive 82/501/EEC, popularly known as the 'Seveso Directive', made risk assessment and reporting obligatory for all EC major industrial installations. Special emphasis was placed on communication with the immediate neighbourhoods so as to avoid panic in case of an accident.

Litigation and the ensuing financial risks played a decisive role in America in the development of the auditing idea. Insurance companies wanted to know the risks they were supposed to cover in environment-related insurance contracts.

Risk assessment and reporting were only the beginning of a long story of environmental auditing. In the early years, the terminology was still fairly loose. There were environmental 'appraisals', 'surveillances', 'surveys', 'reports' or 'reviews'. During the 1980s, the term 'environmental audits' became the standard expression. The EPA published an *Environmental Audit Program* (EPA 130/4–89/001) strongly suggesting to the individual states that they establish environmental auditing rules for all commercial businesses.

Litigation and government action were essential in making the auditing idea a reality. But ultimately it was to a large degree the firms themselves that promoted the eco-audits. They realised that their ecological image was becoming a central element of their commercial success. Consumer groups published ratings of companies for their performance in matters far outside product quality. Social and gender equity, relations with dictatorial regimes in other parts of the world and the environment were the most popular focal points for the rating scores. Maintaining good relations with customers and marketing end-user products successfully became highly dependent on a company's environmental image. A good environmental image, cultivated with sincerity and resting on a solid base of real achievements, became a valuable corporate asset, while loss of environmental credibility could be a permanent handicap.

In the run-up to the 1992 Earth Summit, the International Chamber of Commerce (ICC), based in Paris, published a model for environmental audits that was thought useful for firms throughout the world and susceptible to international harmonisation. The ICC concept foresees a three-step audit: action before the examination, action on the spot and action afterwards. Each step was divided into numerous subroutines. An ecological audit can be a huge undertaking costing millions of dollars for a large firm with numerous sites each audited separately, and employing hundreds of experts inside and outside the firm.

In 1991, the EC adopted a new directive on (voluntary) eco-audits which in the meantime has been implemented by the member countries. Fairly demanding procedures both for the registration of auditors and for the audits themselves were introduced at national levels. Eco-audits are meant to provide exhaustive and useful information which also enables firms to become more efficient with regard to energy and materials use. All this involves substantial costs which cannot easily be shouldered by small firms. On the other hand, a new market has been created for environmental consultancies.

Eco-audits and voluntary agreements (between government and certain branches of industry) have become the favourite instruments of present-day environmental policies; they harmonise with the deregulation trend. But don't be surprised if the scope for real change remains limited so long as prices and incentive structures keep encouraging wasteful use of natural resources.

Chapter 11
We May Have Fifty Years Left to Close the Gaps

In this third part of our book we wanted to convey a sense of urgency for changing the present course of technological and civilisational development. Our intention is not to be doomsdayers, but realistically to assess the speed and momentum of the current trends, including growth rates in China and India, and investors' apparent fascination with those growth phenomena.

Once infrastructures, urban developments and major capital investments worth millions of millions of dollars are in place, they will to a large extent determine subsequent moves. Even readily available potentials of resource efficiency are likely to be ignored if creditors and owners want to see higher returns on their investments (and continue to believe that efficiency will raise rather than lower their costs).

In this book we are not emphasising points of no physical return from various aspects of ecological damage. Rather, the sense of urgency arises from fast-approaching points of *economic* no return, after which it will be very costly to achieve the benefits of resource efficiency. But if we redirect investments and technology development now, we are likely to earn most of the expected benefits at negative cost.

Of course, there are points of physical no return. Extinct species cannot be brought to life again, and major climatic changes will allow for a 'return to normal' only in timespans of thousands of years. Assessing the scientific analyses on climatic change and other ecological menaces, we believe that the world may have some 50 years left to close the gaps that were indicated in Figure 1 or in the discussion of the avalanches of matter.

This final chapter of Part III will be devoted to a more or less quantitative analysis of the chances of closing these gaps. Let us take *The Limits to Growth* as our starting point.

11.1 Beyond the Limits? The Meadowses May Be Right

Donella and Dennis Meadows, together with Jørgen Randers and William Behrens, were the authors of the first major Report to the Club of Rome, *The Limits to Growth*, published in 1972. Some 9 million copies were sold in 29 languages, and the book changed the world. The global civilisation became aware of outer limits that had previously been ignored.

Not surprisingly, it didn't take long for the critics to discover flaws contained, or at least alleged, in the message, in certain details and in the methodology. Poorer countries (and the poor living in rich countries) found it unfair that the rich should declare limits to growth at a time when the poor were just starting to experience economic growth. Resource specialists were able to show that mineral resources, including gas and oil, were far more abundant than the *Limits* authors were assuming (though the authors retorted that this entirely missed the point). Many economists and politicians felt that *Limits* was too pessimistic in general, and John Maddox and his followers in the scientific and engineering communities (see Chapter 10, p 252) said that technological progress had so often produced unexpected answers to any given problem that it was nonsense to speak of rapidly approaching limits.

Indeed, *The Limits to Growth* was based on a deliberately simple computer model, and the results were also very simple. Some of the input data proved wrong. And technology can indeed do fabulous things. *Factor Four* actually wishes to be quoted as witness for this fact. On the other hand, the physical state of the environment has deteriorated to a large extent. The trends of growth have continued and have brought us much closer to certain limits, if perhaps others than those identified in the 1972 study.

The Meadowses and Jørgen Randers got together again 20 years later to work on an update for their book. But they discovered that they would have to write a new book altogether. Too much had changed, say the authors in the preface of *Beyond the Limits* (Meadows *et al.*, 1992, p xiv):

> *As we compiled the numbers, reran the computer model and reflected on what we had learned over two decades, we realised that the passage of time and the continuation of many growth trends had brought the human society to a new position relative to its limits. In 1971 we concluded that the physical limits to human use of materials and energy were somewhere decades ahead. In 1991, when we looked again at the data, the computer model, and our own experience of the world, we realised that in spite of the world's improved technologies, the greater awareness, the stronger environmental policies, many resource and pollution flows had grown beyond sustainable limits.*

For example, between 1970 and 1990, human population has increased from 3.6 to 5.3 thousand million, automobiles from 250 million to 560 million, annual natural gas consumption from 31 to 70 trillion cubic feet and electric generating capacity from 1.1 to 2.6 thousand million kilowatts. Whatever geologists may say about rich and undiscovered resources, consumption cannot continue to grow at this rate on a rounded Earth. Even stabilising consumption at these vast levels is not going to solve the problem. And many analysts say that it's not so much scarce resources but the absorptive capacity of the earth for all the pollutants and wastes that is limiting further growth of resource consumption.

We have come closer to the limits, say the Meadowses: 'Without significant reductions in material and energy flows, there will be in the coming decades an uncontrolled decline in per capita food output, energy use and industrial production' (Meadows *et al.*, p xvi). However, much as they did in *The Limits to Growth* (though hardly anyone read it carefully enough to notice), they themselves also point out the good news: 'The decline is not inevitable. To avoid it, two changes are necessary. The first is a comprehensive revision of policies and practices that perpetuate growth in material consumption and in population. The second is a rapid, drastic increase in the efficiency with which materials and energy are used' (Meadows *et al.*, p xvi).

They are right in saying that efficiency won't be enough. If exponential growth goes on at a rate of 5 per cent per annum, the entire Factor Four efficiency revolution would be eaten up within less than 30 years! Figures 36 and 37 graphically, if schematically, illustrate the effects.

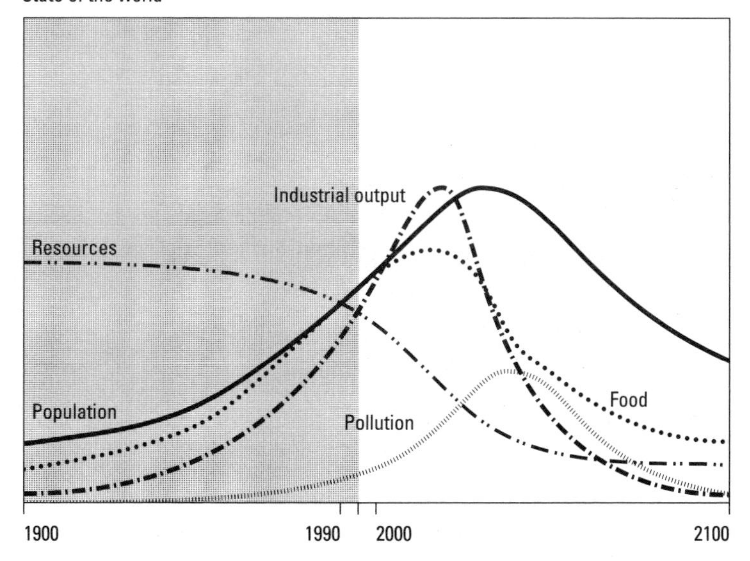

Figure 36 The 'standard run' from The Limits to Growth. *Population and industry output grow until a combination of environmental and natural resource constraints eliminates the capacity of the capital sector to sustain investment (from Meadows* et al., *1992, p 133).*

Several more scenarios have been calculated by the Meadows team. And some scenarios are indeed self-stabilising. The best scenario in the new book is not realistic but has a high heuristic value: it presupposes that sustainability policies (including on population) were implemented as early as 1975. Figure 38 shows the results.

The trouble is that such policies were not in fact introduced in 1975 and are not very likely to be introduced even now. To the contrary, the public attitude, as represented for example by the majority in the 104th US Congress, is not at all pro-environment or pro-sustainability. The authors of *Beyond the Limits* say that '...the transition to a sustainable society requires ... an emphasis on sufficiency, equity, and quality of life rather than on quantity of output. It requires more than productivity and more than technology; it also requires maturity, compassion, and wisdom' (Meadows *et al.*, p xvi). Could these ever be the sentiments of a Newt Gingrich, Prime Minister Matathir of Malaysia or President Cardoso of Brazil?

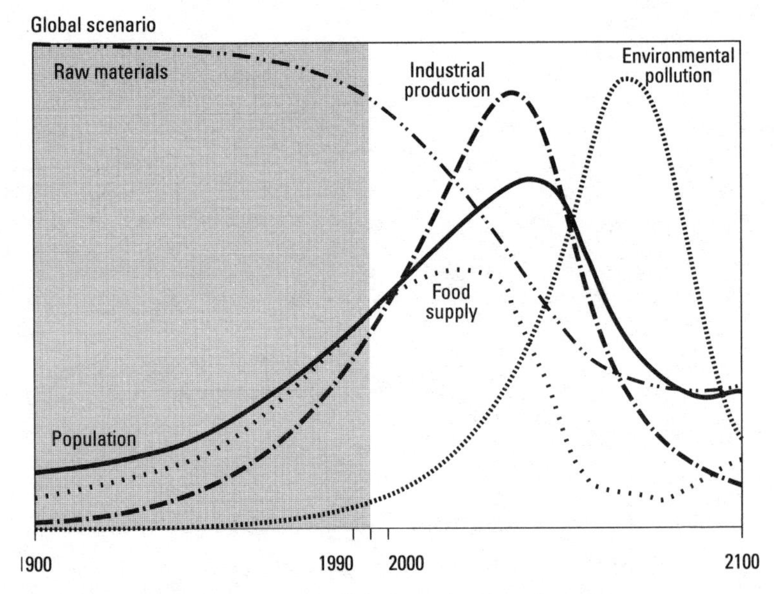

Figure 37 Assuming a doubled stock of exploitable resources, industrial outputs may grow a mere 20 years longer, but otherwise nothing is gained and the fall is all the steeper. (From Meadows et al., *1992, p 135.)*

The Meadowses and Randers give a fairly convincing, if depressing, answer to the question of why politics performs so poorly in preparing for the 21st century. In the graph on page 262, the term of office for political leaders elected in 1990 is juxtaposed to the expected payback period for major corporate investments, the operating life of a power station (the paragon of long-termism in the business world), the lifetime of a child born in 1990, the influence on the ozone layer of CFCs manufactured in 1990 and finally the time horizon of the Meadows group's World3 model.

A fairly fundamental cultural shift and an emphasis on sufficiency may indeed be a way of escaping the destructive dynamics that characterised the dynamic systems depicted in *The Limits to Growth*. But we feel that sufficiency is not something that is promoted by any power dominating the decisions that are being taken in our world. Not even the churches seem to show a consistent line of thinking in this regard. In particular, they often seem to be very hesitant about sufficiency in family planning.

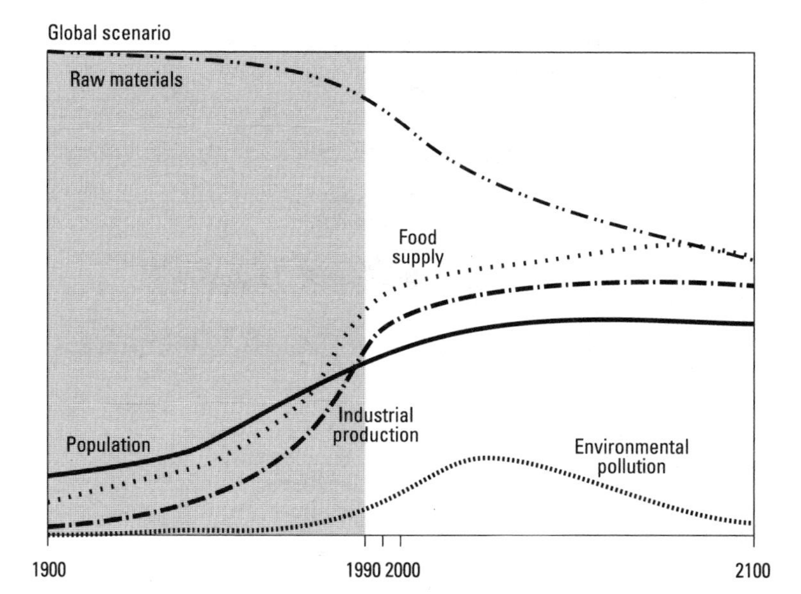

Figure 38 The sustainability scenario presupposes that strict sustainability policies, including on population, were introduced as early as 1975. Even per capita consumption can be kept sustainably at very comfortable levels. (From Meadows et al.*, 1992, p 203.)*

11.2 Population Dynamics

None of the problems discussed at the Earth Summit would be serious if we had only some 500 million people to feed, clothe and shelter. Oddly, however, the population issue was not even a topic for discussion at UNCED. Rumours had it that this was a concession to the Vatican and to certain Islamic countries. It is hard to believe that religious ideas should prevent people and the countries they live in from doing what is needed to preserve God's creation and a decent livelihood for humans on this earth. But there we are. Some religious leaders seem to fear that population policies may lead to a general erosion of respect for life.

World population is increasing at a rate nearing 100 million people per year. The developing countries account for 95 per cent of the increase. On the other hand, every additional US citizen statistically adds more stress to the natural environment than 20 Indians or Bangladeshis. From an ecological point of view, most

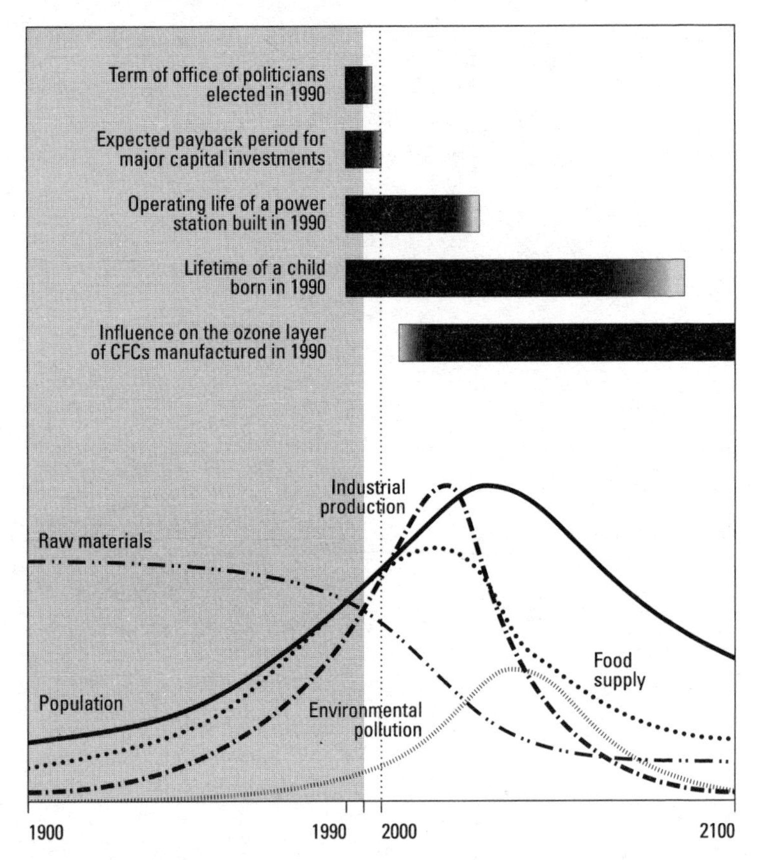

Figure 39 Time horizons of politics, the corporate world, children, the environment and the system dynamics of the World3 model on which Figures 36–38 were based. (From Meadows et al., *1992, p 235.)*

Northern countries are far more overpopulated than India or China. (See the discussion on 'ecological footprints' in Chapter 8, p 220). Correctly, the President's Council on Sustainable Development (PCSD, 1996) put high emphasis on avoiding unwanted births in the US.

A medium estimate by the UN projects total world population at 10 thousand million people by the year 2050, as shown in Figure 40.

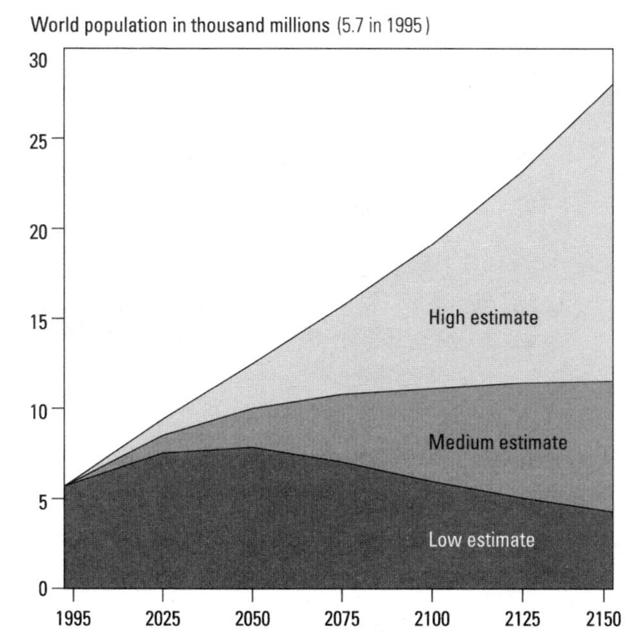

Figure 40 Three variants of population growth during 1995–2150. Even the lowest estimate projects a world population of 8 thousand million for 2050.

Notwithstanding certain political and religious objections, the international community did address population questions. The United Nations convened the ICPD, the International Conference on Population and Development, in Cairo in August 1994. At this conference, the third of its kind, very valuable contributions were made towards rational population policies. In particular, the ICPD emphasised the importance of the role of women, their social status, their education level and their financial independence. Figure 41 from the 1990 UNDP Human Development Report strikingly shows how population growth in ten developing countries correlates with the absence of female education. Can we go as far as suspecting that certain Islamic leaders might be purposely repressing female education with a strategy in mind for a 'demographic victory' of Islam? In a world of limited resources that strategy would also bring poverty and destitution to those same people.

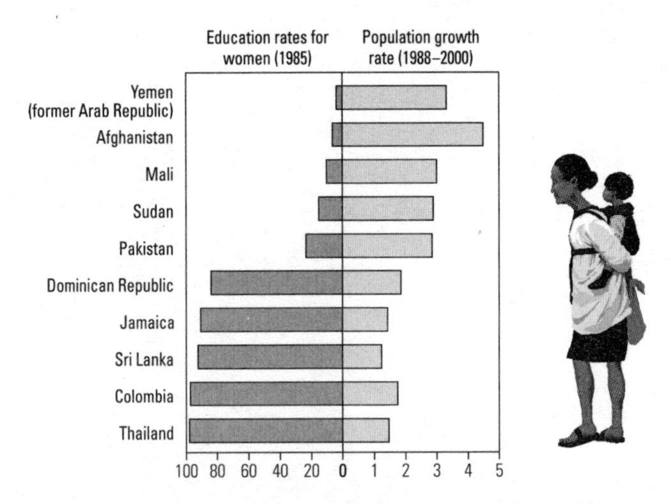

Figure 41 Female education rates (in 1985) showed a striking negative correlation with population increase in ten different developing countries. (From UNDP Human Development Report 1990)

WHAT HAS FACTOR FOUR GOT TO DO WITH POPULATION?

Assuming per capita consumption increases of a mere 1.5 per cent per year (China has maintained annual increases of some 8 per cent for years!), the medium variant would lead to a quadrupling of total consumption from 1995 to 2050. In other words, the total Factor Four revolution – if it took place during the same period – would already be eaten up by this double dynamic of population increase and a very modest increase of per-capita consumption. Nothing of the efficiency revolution would be left for a relief to the overstressed natural environment. How much worse would it be if the efficiency revolution were not taking place!

Politically speaking, the dramatic increase of population would almost inevitably lead to conflicts over land and resources. Migration would reach all continents and countries. It is very much in the interest not only of the poor but also of the rich to halt population growth and eventually to make the lowest scenario of Figure 40 come true.

We ought not to introduce the Factor Four revolution into the population discussion in a merely defensive fashion. It could actu-

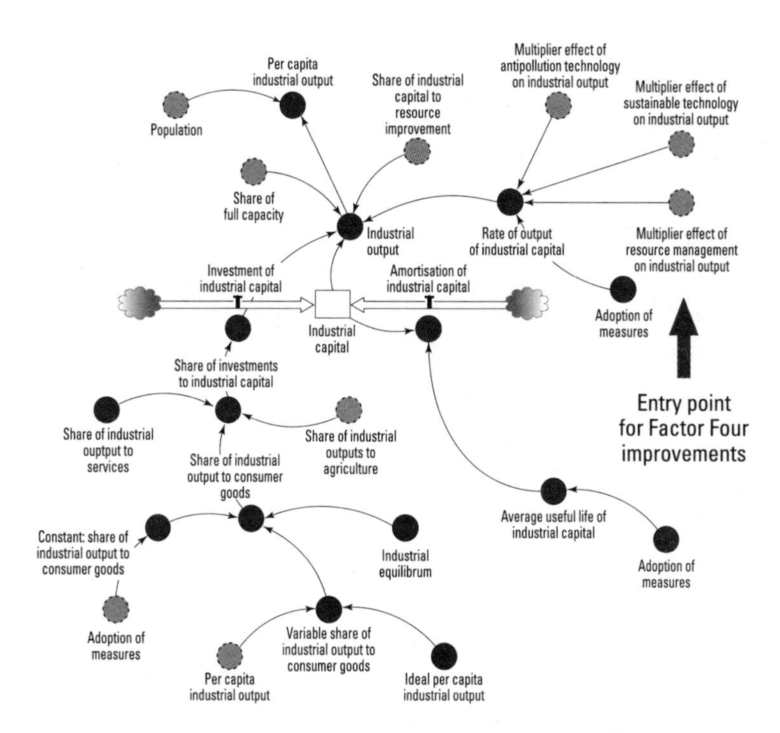

Figure 42 The Industrial Output submodel of the World3/91 model used by Meadows et al. *(1992), p 245. The point of entry for Factor Four improvements is marked with an arrow.*

ally play a decisive role in achieving a stabilisation of world population. It has been a well-established demographic fact for more than 50 years that populations tend to stabilise, almost regardless of religion, when a certain degree of prosperity is reached (which also correlates highly with the independence and self-esteem of women, although that latter correlation is dependent on religion and culture).

If we now learn that American-style prosperity, for ecological reasons, is definitely not conceivable for six or more thousand million people, then the hopes for stabilising population the 'natural' way are very dim. If, however, the efficiency revolution allows prosperity to occur at resource consumption levels roughly a factor of four below America's, we may become hopeful again.

In other words, those in the North who consider population

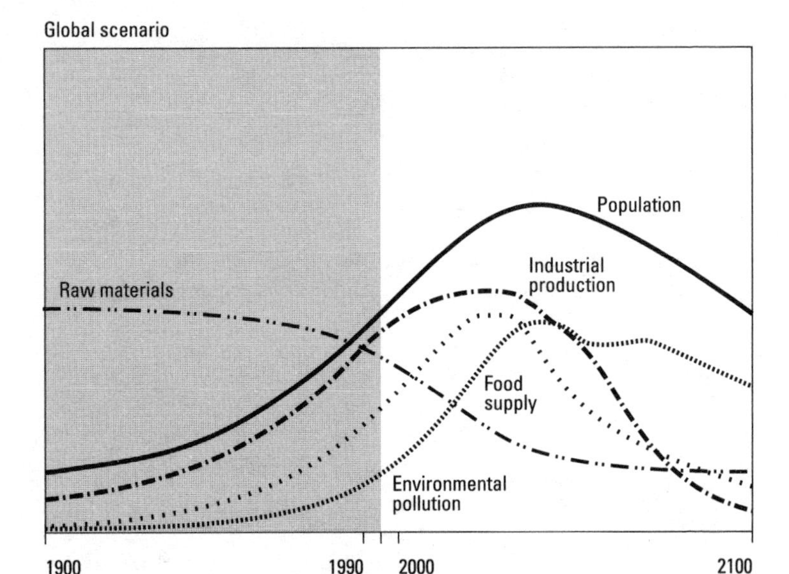

Figure 43 Assuming 2 per cent annual gains in resource productivity, the World3/91 model indicates a stabilisation after 2100 at a fairly low prosperity level – not bad when compared with the wild and catastrophic system behaviour shown in Figures 36 and 37. (The authors gratefully acknowledge Dennis Meadows's willingness to share his software with us, and Harry Lehmann's talent in injecting the Factor Four idea into the World3/91 model.)

growth the biggest menace should do all they can to facilitate both increased prosperity and the efficiency revolution in the South. And as the South is not going to embark on the efficiency revolution on its own, it is all the more urgent for the North to begin the new trend!

11.3 Some Arithmetic about the 21st Century

Chapter 11 has so far shown that conventional wisdom does not provide solutions to the 'limits to growth' challenge, and that population can be stabilised, but not soon, and not without attaining a satisfactory level of wellbeing. After having briefly discussed the

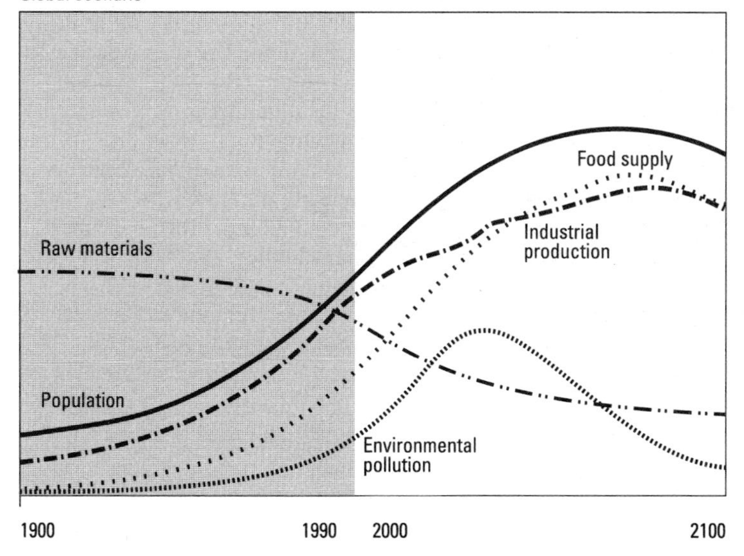

Figure 44 Assuming 4 per cent annual gains in resource productivity, stabilisation can be expected after 2050 at a higher level of prosperity. (Acknowledgements as in Figure 43).

central factor of the Meadowses' system dynamics, population, we may now turn our attention to whether there are scenarios available that move beyond the purely educational value of the 'start sustainability policies in 1975' scenario. Is there a reality-based scenario for a sustainable future?

We address this question by applying the World3/updated 1991 systems dynamics model, which Dennis Meadows has kindly made available to us. Figure 42 gives an impression of the rich complexity of the model. It represents the Industrial Output subsystem only. The full model naturally contains and interconnects a great many other subsystems.

The main idea is to inject the dynamics of the efficiency revolution into the existing model in two simple and fundamental ways. We assume that efficiency gains of 2–4 per cent annually can be obtained (both rather conservative estimates consistent with historic experience; 2 per cent annual gains in, say, energy productivity have been achieved in many societies with little or no policy intervention or deliberate effort). From the prosperity and education parameters we expect a decrease in family size, *i.e.* birth rates decreased by 40 per cent from 2000 to 2100. As a result, a moderately optimistic development is obtained for the 2 per cent gains assumption (Figure 43), and a truly attractive scenario emerges from the 4 per cent gains assumption (Figure 44).

The 21st century need not be depressing at all. If our 'neo-cornucopian' visions come true, even the gravest worldwide distribution problems can be solved without any part of the world's having to accept significant sacrifices in well-being. What cannot be dealt with by globalised computer models are epidemics, not wars and other conflicts. Also, irrational behaviour under the permanent stress of worldwide economic competition cannot be accounted for. We shall be discussing some of these outstanding problems in Part IV.

Part IV
A Brighter Civilisation

We have now looked at:

- an amazing potential for increasing resource productivity (Part I);
- ways and means to unleash the efficiency revolution (Part II); and
- the ecological challenges the earth is facing (Part III).

Where do we stand now?

Efficiency can safely be called a no-regrets strategy. If it is made a high priority in many countries and by businesses worldwide, it will buy a lot of time. But it does not ultimately solve problems. When growth goes on unchecked and when the most profitable opportunities for increasing resource productivity are eventually exhausted, some of the feared ecological and social cataclysms will someday come upon us. We cannot ultimately escape the need for establishing civilisational limits to growth. These must include an end to population growth and an end to the growth of material resource use.

The idea is not, however, to impose in a dictatorial manner any civilisational prescriptions. The idea is rather to arrive at a better understanding of well-being and satisfaction. This better understanding, and moving from understanding to doing, is what we would characterise as a brighter civilisation.

Among the goals are:

- A better understanding of economic growth versus green economics – *i.e.* 'real wealth' or well-being. Some elements of what today is called growth have nothing to do with bottom-line well-being but are just accelerating turnover.
- A better handling of international trade. It is a tragedy that international trade competition is hindering nations world-wide from implementing domestic ecological or social policies they would otherwise willingly adopt.
- A better understanding of noneconomic and nonmaterial values that are indispensable elements of human satisfaction.

These tasks are the central themes of the fourth part of our book. We limit the discussion to three chapters, while recognising that each of these themes deserves a systematic treatment in a book of its own.

Chapter 12 Green Economics

12.1 Green GDP

Two cars pass each other quietly in a country lane. Nothing happens and they contribute little to the GDP. But then one of the drivers, not paying attention, wanders over to the other side of the road and causes a serious accident involving a third approaching car. 'Terrific,' says the GDP: air ambulances, doctors, nurses, breakdown services, car repairs or a new car, legal battles, visits from relations to the injured, compensation for loss of earnings, insurance agents, newspaper reports, tidying up the roadside trees – all these are regarded as formal, professional activities which have to be paid for. Even if no party involved gains any improvement in his or her standard of living and some actually suffer considerable loss, our 'wealth', namely our GDP, still increases (from *Earth Politics*, Weizsäcker, 1994, p 197).

Something is wrong with the popular notion of GDP or GNP (Gross National Product) as an indicator of a country's welfare. To be fair to the GDP, it is not defined as an indicator of wealth or of welfare. It is meant to measure the turnover of economic activities that are susceptible to measurement. A baby's pleasure may be greater at breast-feeding than with a formula bottle, but the bottle and its contents are susceptible to measurement, while the mother's breast-feeding is not; it may even appear negatively in the GDP calculation, as it may prevent the mother from undertaking some measurable professional activity.

The New Economics Foundation (NEF), in conjunction with the Stockholm Environment Institute, has summarised some of the earlier work on more adequate indicators of welfare. The team has used and modified the Index of Sustainable Economic Welfare (ISEW), which was introduced by Clifford W. Cobb as a candidate indicator of well-being in an appendix to the book *For the Common Good* by Herman Daly and John Cobb (1989). (Sometimes, the same indicator is called GPI, or Genuine Progress Indicator.) Contrasting GNP with ISEW reveals that the earlier close correlation between GNP and ISEW collapsed in the mid-1970s (Jackson and Marks, 1994). Since that time, ISEW for the UK, the US and other OECD countries turned downwards while the GNP kept growing, as Figure 45 shows. This goes far toward explaining the seeming paradox of, say, American politics today, that while the official statistics show the economy growing vigorously, most people seem to feel that they are simply running harder to stay in the same place or are even slipping backwards. The explanation is that both views are right, but the official statistics are measuring the wrong thing: most of the growth they report is going to remedial costs, and to bads and nuisances being treated as if they were goods and services, rather than to making people actually better off.

The deeper and more detailed reasons for the growing discrepancy between turnover and *real* wealth are not simple. Quantitatively, the most important factor in distinguishing the two is informal sector wealth (see Chapter 14.2). Other reasons for the discrepancy involve both quantitative and qualitative aspects. GDP does not cover the resource account balance left for future generations. Similarly, environmental degradation appears rather as an advantage in GDP calculations, owing to the economic activities it causes, while it figures negatively in ISEW. Moreover, growing income discrepancies and violence on the street certainly degrade the quality of life for the larger part of the people, even if the average economic turnover goes up. Many such discrepancies were on the rise in OECD countries throughout the 1980s.

Many attempts have been made to adjust GDP or to formulate alternative indicators. Hazel Henderson, in Chapter 6 of her *Paradigms of Progress* (1991), finds rather strong words for what she calls 'the indicator crisis'. More compromising and more comprehensive in scope is Paul Ekins's and Manfred Max-Neef's *Real Life Economics* (1992). These authors shed light on the difficult question of sustainable development.

A relatively conflict-free but effective method was chosen by the statistical office of Germany, namely the creation of 'satellites' to the GDP. Parameters like environmental degradation, unpaid

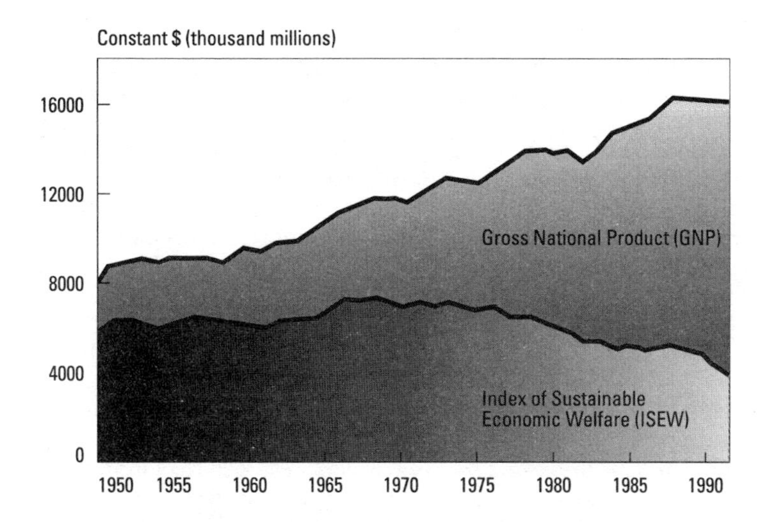

Figure 45 The Index of Sustainable Economic Welfare (ISEW) reflects in a broader sense than GNP how people are faring. In the mid-1970s, the correlation between the two indicators in the US begins to vanish. (After Jackson and Marks, 1994.)

work or educational attainment can be measured independently, but as such they don't intrude on the GDP measurements. The UNDP [United Nations Development Programme]'s *Human Development Index*, comprising education and life expectancy, can be seen as such a satellite.

The ISEW, on the other hand, is not conceived as a system of distant satellites, but rather as a substitute for GNP. This is certainly the ambition in an important new Report to the Club of Rome, *Taking Nature into Account* (van Dieren, 1995). We don't see a need to repeat what is contained in that report. Instead, we should like to address a political difficulty governments will be facing when attempting not only to establish new wealth indicators, but also to act accordingly and to sacrifice some GDP growth to the improvement of the new indicator.

Soon governments (and the indicator academics) will realise that GDP stands for more than just turnover. In the real world of politics, GDP stands chiefly for employment. It thereby also stands for fiscal revenue. In an historic period where addressing unemployment and stable public finance are among the absolutely top political priorities, one should just not expect states or political parties to turn their backs on GDP. Under the present conditions, the lack of GDP growth will be perceived and felt by many as a severe lack of real welfare.

Let us therefore not underestimate the political task of making societies sail towards real and lasting welfare. This task appears to be considerably trickier than the academic task of redefining welfare in a more coherent way. We shall have to look at accompanying features, such as employment, and we have to recognise incentive structures and side-effects which may willy-nilly reconfirm and reinforce classical growth even when the people don't explicitly want it.

FACTOR FOUR FOR A LEANER GDP

What has been said in this book will already help a lot with the political task facing us. It demonstrates that 'growth' can be made much more meaningful again in terms of real wealth. Employment to improve energy efficiency of buildings and to manufacture products with high energy efficiency and long service lives will certainly enhance, not diminish, ISEW. Also, if ecological tax reform (ETR, see Chapter 7) helps to make human labour more affordable again for employers, while putting economic pressure on resource use, it will be easier even in the unemployment-stricken European countries to regain some sort of satisfactory employment levels. Both efficiency and ETR will help in re-establishing a reasonable degree of correlation between GNP and ISEW.

In a transitional period, which may last as long as 50 years, the initial investments for a resource-efficient civilisation may actually keep employment rates rather high. The replacement of inefficient stock by increasingly efficient appliances, cars, homes and infrastructures will inevitably take many decades. And the pioneering countries are likely to enjoy a boost in their exports. From an ISEW point of view all these new jobs will enhance rather than diminish welfare.

However, all this will not ultimately solve the fundamental problem that turnover is not the same as wealth. Accidents, crime, environmental degradation and the divide between rich and poor keep real wealth lagging behind economic turnover. Moreover, more and more human labour, including white-collar work, will probably be done by machines, even if resources become much more expensive and human labour becomes more affordable again for the employers. More jobs will be lost. Some useful new jobs can be created in their place.

But if material productivity in particular drastically rises, the manufacturing of new goods will not expand but shrink. Economic value will lie in stock rather than in turnover.

12.2 The Service Economy

The more we think in terms of end-user satisfaction, the more we comprehend production processes as the early parts of service processes. The 'services' of providing agreeable warmth, illumination, [taste] satiation, entertainment, security and so on should dominate the production philosophy, not the other way round. This has been the starting point for Orio Giarini and Walter Stahel (1993) in their considerations both of material productivity (see the introduction to Chapter 2) and of what they call the *service economy*. Orio Giarini, together with Patrick Liedtke, is preparing a new Report to the Club of Rome on the future of work, to appear in 1997, further elaborating this important trend.

The service economy can also be labelled the '*stock economy*' as opposed to the flow- or turnover-economy. It will be the services rendered, using the existing stock of goods, that will serve as the yardstick of success.

An important element of the service economy is '*tertiarisation of the secondary sector*', meaning to shift emphasis in the manufacturing (secondary) sector away from production and towards consumer satisfaction by guaranteeing tailor-made products and first-class service and maintenance.

Given the urgency of saving and creating jobs, political talent will be needed to communicate the desirability of not expanding the stock of goods. Although a stagnating stock of goods sounds like a nightmare for politicians and industrialists of today, the prospects of success in job creation and job-savings are actually not that bad.

One of the best strategies in this regard happens to be at the heart of resource productivity. Resource productivity means skilled labour to satisfy clients, to maintain and repair machines, to recover and recycle used materials and so on. Some of the salaries for the new service people will be paid from the money value of the resource savings. On the other hand, some jobs will also be lost, *e.g.* in mining, power generation and the shipping of raw materials and goods. The nature of jobs will certainly keep changing and the percentage of service jobs will keep increasing.

For the North, worried by floods of cheap and short-lived consumer goods from the South, the service economy may bring some relief with regard to jobs. Maintenance, repair and re-manufacturing of and financial services for autos used in the US or Europe will, as a rule, be done in the region, much of it in town, even if the car itself may come from Korea or Mexico. We believe that appreciably more jobs will be saved or created by moving actively towards the service economy than by a laissez-faire continuation of the old industrial paradigm. However, as Walter Stahel pointed out in a personal communication, no assessment of the job balance has ever been made. We may add that it would be fairly difficult methodologically to ascribe the creation and loss of jobs directly to any specific policy.

12.3 Working until the Age of Eighty?

Geneviève Reday-Mulvey, who works with Giarini and Stahel, has made an important observation from which she derives an equally important proposal. She says that the service economy is also offering more job opportunities for people with lifelong experience. Knowing an industry's customers may, in many cases, count for more than knowing myriad details of the laws of physics. She combines this fact with an urgent social policy question facing all industrialised countries: how can we guarantee pensions for the growing portion of older people in our societies? Moreover, many of these elderly people are much healthier and more functional than people of that age were 50 or 100 years ago. From these observations, Reday-Mulvey derives the idea of nonobligatory but remunerative part-time work done by the elderly. Plate 15 symbolises the mutual relations among demographics, the changing quality of work, a new life cycle (in which healthy, active living extends well into old age) and the problem of financing tomorrow's pensions.

In the context of suggesting ways to help solve a worsening unemployment problem, presenting ideas for people to stay longer in their jobs may seem counterintuitive – 'taking jobs away from the younger ones.' In reality, it is the other way around. If an ever heavier burden of pension payments has to be shouldered by the formal jobs of the young, then the price of the latter goes up absurdly. Jobs will price themselves out of the market (as is already the case in several European countries). Moreover, the type of jobs offered to and accepted by the elderly will typically be in a domain that is unattractive for most youngsters: mostly personal care jobs with still older people, or repair jobs for 'outdated' goods and appliances, or jobs requiring decades of experience.

Reducing the financial burden from pensions may become imperative anyway. The service economy, on the other hand, could make this a highly attractive and profitable operation for the whole economy. Finally, it can be assumed that the prospect for many clients of obtaining affordable and reliable services will induce them to divert appreciable amounts of additional money to paying for such services from what they would otherwise spend on activities with little effect on domestic jobs.

Chapter 13
Trade and the Environment

13.1 Free Trade Strengthens Capital and Weakens Labour and the Environment

International trade and competition are meant to improve the economic well-being of people. At the same time, international competition is being invoked in all countries as an explanation and excuse for undoing social policy and environmental regulation, for shedding labour and for impatience to adopt technologies that put the environment or human health at risk.

The assertions by economists and politicians that the benefits of trade are greater than risks and costs find positive resonance with those owning an appreciable amount of capital. To them, free trade offers opportunities to let their money work where it fetches the highest returns. Those who don't have money find it more difficult to see the benefits of free capital movements, and often feel blackmailed by the capital owner's threat to leave the country.

The threat to leave the country also applies to any attempt by politicians to tax capital. In earlier days, taxing capital was one of the most natural ways to augment the public purse – provided it was kept at prudent levels. It always had a legitimate redistributive effect. In times of high unemployment, taxing capital would seem even more reasonable, because taxing labour has become particularly problematic, and because growing social discrepancies make redistribution ever more legitimate. However, capital can easily evade higher taxes by emigrating. Labour cannot. Reflecting the divergent development of their respective bargaining powers,

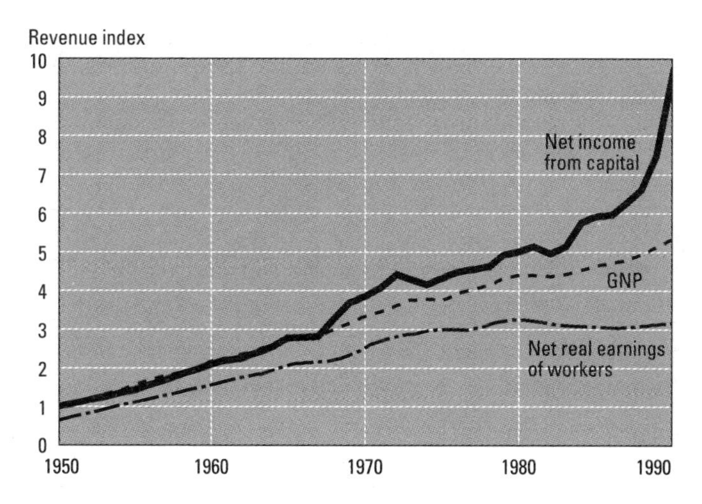

Figure 46 Until around 1980, income from both labour and capital went more or less in parallel with German GDP growth. Dramatic changes have occurred since that time. Workers' income has stagnated, while income from capital has exploded. (Redesigned after Afheldt, 1994, pp 30–31.)

labour and capital income levels have gone in strikingly different directions since the early 1980s, as Figure 46 documents for one OECD country, the Federal Republic of Germany. In the US, workers' earnings even went appreciably downwards as income from capital soared.

The turn from the 1970s to the 1980s marked the beginning of a sweeping victory of the free-market philosophy combined with 'restructuring' of social welfare systems. It is not far-fetched to assume that there has been a causal relation between the two.

'Competition Is War'

More alarming than divergent income trends is the danger of collapse of our civilisational morale. There are now serious economists who praise the destructive force of markets, with some in the profession eagerly taking up Joseph Schumpeter's concept of 'creative destruction'. It is noteworthy that, when Schumpeter was writing, the destructive power of markets was trivial compared with the harm they can do today. In *Hypercompetition* (1994), Richard D'Aveni goes a lot further than Schumpeter, saying that 'competition is war.' To him, the goal of competition is not to be better than your rival, but to destroy him. There's no place left for ethics in this grim philosophy.

Some seek consolation in the rather theoretical assumption that this 'war' is between speculative traders, or impersonal capital owners such as pension funds, involved in computerised movements of capital around the globe. But there are always people linked in some way to the capital. Let the losers become angry about the quasi-military attitude of the winners, and we will be in the midst of a new type of Cold War that can only too easily become linked to armaments, blackmail and more.

Thinking of the consequences for civilisation of the 'competition is war' concept, we hope that it will remain only a transitory fashion among some extremists in the economics profession.

CAN WE STILL AFFORD SOCIAL POLICY?

Completely unrestricted markets seem to have become a distinct danger for social policy. It is ironic – is it not? – that on the heels of a period of steady and massive progress in welfare creation, of freer access to the world's resources and of significant disarmament following the sudden end of the Cold War, the prosperous countries can no longer 'afford' social policy. But this is precisely what industry and many economists are now trying to tell the public. Complaints about this blatant discrepancy between common sense and economic rhetoric was the refrain of many speeches at the Copenhagen World Summit for Social Development in March 1995.

To counter such complaints, free-market advocates tend to reassure the public and the losers that free markets do increase overall welfare and help keep prices for imported goods at levels much below anything protectionist strategies would be able to maintain. Moreover, they claim that it is good for America (or Britain or Germany) if its capital is allowed to move freely around the world to fetch the highest returns possible, and that taxing capital was generally counterproductive in terms of economic growth. But then, if the unemployed and the poor hear that the prices for fruit, textiles and cameras will be much higher without trade, they tend to react by saying they would be quite happy without strawberries in winter and with costlier cameras, as long as they were enjoying a reasonably secure job – and a healthy environment too.

New Protectionism?

Sir James Goldsmith (1995), one of the most successful businessmen in the free-market world, has become an eloquent critic of unrestricted trade for the reasons indicated above. He says that since about 1990, with the entry into the world market of nearly 4 thousand million people accepting wages easily 20-fold below typical OECD salaries, it has become simply impossible to create secure jobs in Western Europe and other OECD areas. He challenges the assertion by free-market defenders of goods becoming cheaper this way, citing the case of Nike, the shoe manufacturer which relocated most of its fabrication from the US to Asia –with no effect on shoe prices but much higher profits to the company.

Goldsmith sees no overall benefits in exchange for the corresponding losses, and ends up advocating regional protectionism. Regions, of course, would be free to negotiate about what they want to exchange for mutual benefit. Goldsmith's book has been the number-one best seller in France for months and had a major impact on some quarters of the political spectrum of continental Europe. However, when he founded the Referendum Party in the UK, with its anti-Maastricht platform, many of his earlier friends turned their backs on him. Old-fashioned protectionism and national economies simply seem out-of-step with our times. Echoes of his call for regional protectionism can be heard in the US, as they were, for example, in the campaign speeches of Pat Buchanan in the 1996 Republican primaries.

In response to neo-protectionist sentiments, the World Bank devoted its 1995 Development Report to the issue of jobs in a free-market economy. The Bank argued that the share of goods imported into OECD countries from low-wage areas is too small to make any difference to the wage balance, and that low-price imported goods keep inflation down and thus help to prevent a price-wage spiral in the OECD. In what can only be seen as cynical obfuscation, given labour's decreasing bargaining power, the report stresses the importance at national levels of the trade unions in negotiating and securing high salaries and adequate social security for the workforce.

In the end, however, the report's authors cannot avoid recognising that unemployment is on the rise in the entire OECD area, with the exception of the US, where a sufficient number of people are accepting jobs at a level of earnings that is becoming comparable, in terms of purchasing power, to that in developing countries. So, empirically, Goldsmith seems closer to reality than the textbook-oriented economists writing for the Bank.

What is more, the entire World Bank report fails to mention the environment. It ignores the fact that the desperate (and partly successful) attempts by OECD governments to compensate job losses by state interventionist job creation schemes almost invariably lead to further physical expansion. The favourite job-maker is still expansion of the transport infrastructure – sheer madness given the massive growth of car use in the developed world and all the associated environmental and logistical problems. In Germany alone, there are already four times more cars on the roads than there are in all of Africa! All we have said in the context of sustainable development is in direct contradiction to these frenetic governmental programmes of job creation.

Given all the destructive potential of free trade, however, it is no trivial matter to come up with adequate alternatives. Yet if we don't want to subscribe to the protectionist philosophy of Sir James Goldsmith, the least we ought to do is to understand the conditions under which the negative environmental effects of free trade can be minimised.

CONDITIONS FOR THE FUNCTIONING OF THE MARKET

As a matter of fact, economic theory itself entails important reservations about the superiority of free trade. In his seminal book, *Trade Policy and Economic Welfare* (1974, p 8), W. Max Corden stressed that 'Theory does not "say" – as is often asserted by the ill-informed or badly taught – that "free trade is best." It says that, *given certain assumptions*, it is "best".'

Prominent among these assumptions is the smooth functioning of the price mechanism. Economists Paul Ekins, Carl Folke and Robert Costanza (1994) on whom we rely in this passage, make the point that trade is distorted, since prices do not reflect the full cost of production. Much less are prices reflecting the hypothetical costs of resource depletion and of environmental degradation. Mining the environment leads to prices' deceiving both producers and customers about true costs. During the 1970s and 1980s, at a time of steadily rising resource consumption, market prices for natural resources fell drastically, signalling to consumers that more, not less, was available.

DEVELOPMENT TO DEPLETION?

When the Ivory Coast in the two decades following independence sacrificed much its natural treasures to the production of cash crops and other export commodities, the young nation became a hero of the international banking community. Here was a country 'taking off', enjoying a stable currency (linked to the French franc) and a 'stable' political climate. Well, soon enough, the party was over. What has remained are skyscrapers, fancy hotels and an élite habituated to Western consumption styles, but otherwise widespread destitution, a devastated natural environment and political instability. Neighbouring Ghana, the current darling of international financial institutions, is going down the same track. The Solomon Islands, much smaller than Ghana and the Ivory Coast, destroyed their forests at such a rate that even the IMF (International Monetary Fund) became nervous and admonished the country to adopt a more cautious pace.

We hardly need further instruction about what is unsustainable.

As we write this, living in prosperous countries of the North, we hasten to say that the North has done more than its share in destroying natural resources at home and abroad, and that it was to an embarrassing extent Northern firms, consultants, banks, governments and ideologies that inspired or executed the destruction in these developing countries.

Not that any sinister clan was conspiring to plan and execute the drama of destruction. No, it was just the prevalent philosophy of leaving it all to the markets. This implies that much was left to entrepreneurs and consultants, who were rewarded by the international markets for overexploiting the land. Annual profits were simply higher for unsustainable resource exploitation than for sustainable methods. Clearly, prices under the conditions of international competition were telling the exploiters, both locally and internationally, that they could make sufficient profits only when they were destroying the resource base.

From these observations readers should not conclude that bureaucratic socialism, national protectionism or other ideologies impeding the free flow of goods, let alone information, would help the environment. In fact, as Stephan Schmidheiny (1992) convincingly emphasises, free markets can greatly help the spread of environmentally benign methods and technologies. But something has to be done to reconcile better global environmental protection and the principles of free trade.

13.2 Can the WTO Become Green?

International trade was not invented to protect the environment. When world trade advanced to the forefront of international debate, shortly after World War II, the environment was simply not an issue. The General Agreement on Tariffs and Trade, GATT, was created in 1947 and has since been amended and fortified in eight major rounds, the last one, the 'Uruguay Round', being the most important. Begun in 1986, it came to its conclusion in a grand ceremonial event in Marrakech, Morocco, in April 1994. A major outcome was the creation of a powerful international organisation, the World Trade Organisation (WTO), to replace GATT.

Meanwhile, the environment had for years been a major international concern, as witnessed by the largest diplomatic event in this century, the 1992 Earth Summit of Rio de Janeiro, which attracted more than a hundred heads of state and of government. And yet neither the Earth Summit nor the Uruguay Round addressed the issue of trade and the environment.

Environmentalist Michael Northrop called the Uruguay Round a GATTastrophe and showed that, through the new procedures, it was becoming exceedingly difficult for environmental and consumer groups to intervene in trade issues. And the then-GATT negotiator for the US, Clayton Yeutter, was overheard in 1990 saying that one of his goals at the Uruguay Round was to overturn health and environmental regulations being adopted by Congress.

Governments and free-trade advocates don't usually see any serious conflict arising between free trade and the environment. Just in time for the Earth Summit, the GATT Secretariat published a paper, *Trade and the Environment*, which plainly says that expanding trade is good for the environment – on the theory that trade makes us richer, so we can afford to use a growing proportion of national expenditure for the environment. The GATT paper attributes all responsibilities for the environment to governments and sees trade only as a 'magnifier' of national policies: 'If the policies necessary for sustainable development are in place, trade promotes development that is sustainable... Clearly, the correct action ... is to work for the adoption of an appropriate overall *domestic* environmental policy, rather than focusing attention on problems that are allegedly trade-related' (emphasis added).

The study emphasises that GATT rules place essentially no constraints on the ability of countries to use appropriate policies to protect their environment from damage from domestic production activities or from the consumption of domestically produced or imported products.

All this is legalistic rhetoric. Except for passing references legitimising measures to protect living things or exhaustible resources, GATT simply doesn't talk about the environment and so, it might be claimed, doesn't place 'constraints' on domestic environmental policies. But even in strictly legal terms, the claim of noninterference has not been valid since 1991, when a GATT panel consisting of three GATT insiders ruled against the US in the famous tuna–dolphin case. The US had imposed import restrictions against tuna meat from Mexico, Venezuela and other countries whose catching methods regularly kill dolphins. Lori Wallach, staff attorney at Public Citizen's Congress Watch, commented: 'This case is the smoking gun. We have actually seen GATT declare that a US environmental law [the 1972 Marine Mammal Protection Act] must go.'

Much more relevant and much more worrying than the legal side of GATT's seemingly benign language is the actual effect of trade liberalisation on economic behaviour in all countries. Europe has seen what happened after the adoption in 1987 of the Single European Act, with its Four Freedoms, and the US experienced something similar in the NAFTA (North American Free Trade Agreement) context five years later. It began in the business community with anxieties regarding competitiveness. The anxieties proved justified in all countries concerned. The only solution for businesses was aggressive labour 'rationalisation'. When massive unemployment ensued, environmental issues virtually disappeared from the public agenda. A legal rollback, not only deregulation, took place. If environmental activists and politicians in European countries today were encouraged by WTO officials to domestically 'adopt policies necessary for sustainable development', there would be a chorus of outrage at such hypocrisy.

Of course, free-trade defenders will argue that Europe should be happy, having regained through the Single Market some degree of competitiveness in time for the inescapable trade battles with the Pacific Rim states. This may actually be true. But it's a purely economic statement with no consolation value whatsoever to those who are concerned about the European or the Asiatic or the global environment.

There are attempts to reconcile trade and environment by international agreements on standards or by a more generous interpretation of WTO's GATT Articles XXb (legitimising measures 'necessary to protect human, animal or plant life or health') and XXg (legitimising measures 'relating to the conservation of exhaustible natural resources'). But what can be expected from international standards if even the European Union, with its Community-wide adopted standards and its elaborated legal apparatus, is virtually unable to enforce its directives? Prospects for effectively greening the WTO (or NAFTA or the EU, for that matter) seem rather dim. The WTO ministerial meeting on Trade and Environment, held in Singapore, December 1996, was the first ever GATT/WTO conference explicitly addressing trade and the environment. But nothing came out of it that could even remotely change the pessimistic assessment above.

13.3 A Role for *Factor Four* in Trade and the Environment

Harmonisation with Economic Gravitation, not Against

Many free-trade problems would be greatly alleviated if environmental regulations were valid (and strictly observed) globally. But the trouble with international harmonisation of environmental policies has always been that all those policies appeared as costs and hence as a drag on competitiveness.

We believe that the picture could fundamentally change if it were possible to transform sustainable development into a competitive advantage – analogous, for example, to microelectronics. To internationalise microelectronics, no quarrelsome and prolonged international harmonisation conferences were needed. The new technology spread on its own. It went with the economic 'gravitation', not against it.

We do believe, as we said in Parts I and II of this book, that the efficiency revolution is to a large extent profitable for a country. It would provide competitive advantages to those countries pioneering it. For the others, it would be dangerous to miss the boat. We also said that some elements of the efficiency revolution are profitable now at the company level. This was confirmed by a new study (WBCSD, 1997) giving a picture of added shareholder value for firms emphasising eco-efficiency. But we emphasised that the state can do much to expand dramatically the range of profitability for both producers and consumers. Under the foreseeable conditions of added stress on limited resources and on the environment, the profitability of efficiency is bound to increase further. At any rate, Factor Four appears more likely than, say, dolphin protection or SO_2 abatement to spread with economic gravitation.

Is Efficiency Policy Compatible with WTO?

Along with the potential competitive advantages, however, some real WTO problems could arise if individual states or groups of states decided to make the efficiency revolution a high domestic priority:

- How would the WTO react if any country adopted ambitious auto-efficiency standards, *de facto* banning the import of vehicles not meeting the standards? (Some European automakers have already challenged the mild US CAFE standards on precisely this ground even though those standards are manifestly nondiscriminatory.)
- How would the WTO react if any country introduced an ecological tax reform and, rather than giving exemption to their energy-intensive industries, introduced a tariff on 'grey energy'? (see Chapter 7, p 206)
- How would the WTO react to an import ban by any country on meat or tomatoes involving what may be defined as excessive energy consumption?
- How would the WTO react to a country's demonstrating a very high energy efficiency in its old-fashioned farming methods, and using this for banning or impeding the introduction of high-tech seeds that may involve energy-intensive agriculture?
- How would the WTO react to any tax on primary raw materials which offers a competitive advantage to secondary raw materials domestically won from waste?

Theoretically, much could be legitimised by GATT Article XXg. But if the intention is to conserve global and not specifically domestic resources, it may be difficult to argue for the measure in terms of the legal language of the WTO.

Here we may have very good cause for two GATT/WTO amendments not requiring a big new Round:

- The protection of global resources (or the global commons) should deserve no less support than the protection of national resources and should therefore be granted equivalent status under Article XXg. Such an amendment could be related to Principle 7 of the Rio Declaration, which all WTO member states have signed. Principle 7 says that 'States shall co-operate in a spirit of global partnership to conserve, protect and restore the health and integrity of the Earth's ecosystem.'
- Article XXb should be amended to include exemptions in favour of environmentally sound production methods (not only products). A precedent to this is the Montréal Protocol, which bans the use of CFCs in production processes. (It is still to be feared that a WTO Panel may decide against the validity of the Montréal Protocol.) Again, the Rio Declaration may be

> used in favour of the amendment: Rio Principle 14 says that 'States shall effectively co-operate to discourage or prevent the relocation and transfer to other States of any action or substance that causes severe environmental degradation.'

Another legitimate amendment to the WTO rules would be to admit environment advocates to the WTO panels and to have publicly accessible meetings of these panels. Without public access there can of course be no accountability, and without accountability one can expect the logical consequences of traders making the trading rules.

A much more radical proposal was made by Edith Brown Weiss (1992). In view of the inactivity for more than 15 years of the GATT's Trade and Environment Committee, and the notorious imbalance of all rulings in favour of trade and against environmental protection, she is asking for a new and common legal framework for free trade and the environment.

REDUCING SUBSIDIES

Let us return to more conventional approaches and to a concrete example. The principle enunciated in Chapter 7 – prices telling the truth – should be uncompromisingly applied to the entire transport sector. In the present world, most states give massive open and hidden state subsidies to the transport, industrial and farm sectors. That may have been legitimate in a world of low mobility, and it may still be an acceptable policy of international lending institutions helping the least developed countries. In general, however, such subsidies should no longer be given.

We are not talking about trivial amounts. According to a study for the Earth Council (de Moor, 1997), ecologically damaging subsidies worth some $700 thousand million per year are given worldwide to the four sectors of auto, energy, water and agriculture.

If external costs of transportation were fully taken into account, the hidden subsidies would be much larger still. External costs are defined, after all, as costs shouldered by the general public rather than by those who created the burdens in the first place. Some respectable estimates suggest that total US distortions in transport's apparent price could approach $700 thousand million per year.

The methods of applying the true costs to hauliers may include road pricing, taxes on air traffic fuels, cost-covering insurance for sea and air travel, and ecological tax reform (Chapter 7). By a package of such measures, the road transport prices per tonne-kilometre could legitimately be increased substantially. Similarly, prices for air and sea travel can be legitimately elevated to reflect the true costs. A somewhat greater mutual 'separation' between competitors would result, without any tariff or red-tape protectionism.

This soft decoupling, together with consumer preferences for small-distance produce (cf. Chapter 3, pp 117–121), could improve the competitive position of local products against products coming long distances and presently achieving their apparently low prices through artificially low transport costs or other environmental neglect. Also, the full potential of electronic travel (pp 112–117) can be harnessed if physical transport becomes costlier.

Because they are perfectly in line with the general philosophy of free trade, and yet are appealing to conservationists and to the defenders of local farming and marketing, our ideas may serve as a contribution to solving certain trade and environment conflicts.

13.4 A Philosophical Point about Trade and Darwinism

The principles of the market economy and of international trade find their philosophical roots in Adam Smith's *The Wealth of Nations* and in David Ricardo's concepts of specialisation and the competitive advantages of nations. This combined theory asserted that the pursuit of self-interest and international competition will eventually help us all by increasing the total wealth available for allocation and distribution.

A rather similar hypothesis was put forward by Charles Darwin, nearly a century after Adam Smith, to explain the mechanisms of biological evolution. The theory of evolution provided a grand account of the astonishing unfolding of biodiversity and higher development from primitive to complex life forms in the course of millions of years. Yet the survival of the fittest is about the only element of Darwin's theory of which the broader public has any awareness.

'Darwinism' and the survival of the fittest is always on the minds of economists when they think of competition, technological progress and international trade. Of course, after the criminal abuses of [social] Darwinism by the German Nazi regime, they are cautious about explicit analogies between Darwinism and social and economic development. But the economic doctrine of legitimate victories of the economically successful over weaker competitors is undeniably related to Darwinism. Joseph Schumpeter's notion of 'creative destruction', which has often been cited in the context of 'restructuring' the former Communist countries, involves a quite direct use of Darwinian language. (Schumpeter developed the core of these ideas before the Nazi era and before emigrating to the US.)

EVOLUTION IS THE INCREASE OF DIVERSITY, NOT ITS DESTRUCTION

The philosophical point we want to make is that all this is a distorted understanding of Charles Darwin. To put it simply: Darwin was describing and explaining the increase of diversity during evolution. Simplistic economic Darwinism, in contrast, is about the destruction of diversity.

Instinctively, we assume that trade and the working of the market mechanisms lead to ever-increasing diversity. We see it daily in our supermarkets, don't we? There are mangoes from Brazil, watches from Hong Kong, silk ties from Italy and Swiss chocolates made from Ghanaian cocoa beans. Economic competition has brought advanced technology and an ever-increasing diversity of options into the High Street shops.

The fact of the matter is, however, that diversity is not always on the winning side. How many local beverages disappeared in favour of Coca-Cola? How many fruit and vegetable varieties were lost in the course of market standardisation? According to a study of the Rural Advancement Fund International, quoted by Fowler and Mooney (1990, pp 61–67), some 97 per cent of the vegetable varieties recorded in 1903 have since been lost. Of the 35 rhubarb varieties available in 1903 only one is left and is stored in the US National Seed Storage Laboratory (NSSL). With apples, the situation is slightly better. Of the 7,098 varieties in use in the 19th century, 'only' 6,121 or 86 per cent were lost.

What was lost was not so much the colourful beauty or the size of apples or vegetables. To the contrary, the conspicuous properties improved in the average selection of produce at the greengrocer's. What was mostly lost were vitamins, variation of taste, spread over the growing seasons and climates, suitability for a wider range of soils and robustness against pests.

Marketing needs (not really consumer preferences) call for a reduction of varieties. Cost-effective marketing campaigns depend on having millions of copies of a few identical products. Similarly, industrial processing is not economically feasible for specialities of a few thousand items. Put in Darwinian language, only a few varieties of cucumbers, rhubarb or pears can survive under the imperative of the economies of scale.

Isn't that strange? Wasn't the struggle for survival Darwin's explanation for the increase of biodiversity over the aeons? Indeed, his *Origin of Species by Means of Natural Selection* (1859) pinpointed the steady interplay between variations and selection, and showed with ample empirical evidence that this interplay should lead to ever growing specialisation, diversity and optimised (mostly sustainable) use of the scarce resource base.

The Creative Force of Isolation

The most striking proof for Darwin's theory he found in insular habitats such as the Galápagos Islands. The finches he first described were quite distinct from those anywhere else in the world. In the absence from the islands of woodpeckers, some finches learned to pick insects from inside plants using cactus spines as tools (Plate 15). In the absence of parrots, some developed very strong beaks. In the absence of vampire bats, some even learned to suck blood from warm-blooded animals. None of these abilities is very finchlike, but all are quite creative and enhance diversity.

Modern economic theory has a dislike for insular conditions. It is the nature of unrestricted trade to intrude everywhere and to crush all barriers that may still persist. Indeed, 'removing barriers' is the favourite term of the free market's most ardent defenders. Diversity is not a high priority for economic theory. It keeps its shadowy place in antitrust laws, but is easily sacrificed if the 'laws' of international competition demand it.

Let us caricature the conflict between traditional economics and diversity. Let us imagine Charles Darwin had not been a 19th-century naturalist, but a modern economist stranded on the Galápagos Islands. Immediately, he would have demanded that a land bridge be built between Ecuador and the islands to 'remove the barriers' to woodpeckers' conquering the islands and eradicating the lamentably inefficient spine-using finches. And economist Darwin would go on to say that this would be good for all and for economic evolution and development.

Of course, it's an unfair caricature. But it may show that a philosophical conflict remains between the principle of unrestricted trade and the protection and creation of diversity. As diversity both in biology and in social systems stands for options and for resilience, we feel that it needs protection. But we cannot in this short book offer a broader discussion of policies to protect diversity in a competitive environment.

Chapter 14
Nonmaterial Wealth

14.1 Insatiable Consumption May Outpace the Efficiency Revolution

Numerous proofs were given in this book that gross material turnovers are very inadequate indicators of wellbeing. Chapter 12 also reminded us that the GDP does not measure well-being, but only turnovers insofar as they become visible in monetary terms. The rising discrepancy between GDP and 'real wealth' was illustrated in Figure 45. And in Chapter 11 we argued that efficiency will essentially serve to buy time. We urgently need to make good use of that time to develop a civilisation that is truly sustainable. This, we believe, is the idea behind Principle 8 (on sustainable consumption and lifestyles) of the Rio declaration. We won't have addressed this challenge adequately unless we also say something about nonmaterial aspects of wellbeing.

Rapid progress should not be expected in a world in which both the material basis of survival and social status are inextricably connected with monetary income drawn from capital or from a formal job. We believe that much of the non-material aspects of wellbeing has to do with satisfaction experienced outside the world of paid employment and outside the money economy.

Look at an unhappy child and look at its mother coming to comfort it. The child's satisfaction is likely to be greater than most satisfactions experienced in the consumption of marketed goods and services. But to an economist, the child's satisfaction and the mother's comforting don't count. They do not appear in the GDP. Economists have a tendency to recognise needs or wants only if they can be fulfilled by goods and services that are sold in the marketplace. Our society has a similar tendency of recognising only the economically active people who make their living from commercially satisfying the needs of others. Many of these economically active people make their living from successfully transforming any conceivable dissatisfaction into a need that can be fulfilled by whatever good or service they are offering in the marketplace.

Since the invention of modern advertising and marketing, needs are even being actively aroused and created. Thousands of academically trained psychologists make their living in the advertising industry or in marketing departments by helping to 'create' needs. The economy's 'success' in terms of GDP and full employment hinges a lot on the successful work of these psychologists and their business partners.

So let us not be surprised that many dissatisfactions are becoming needs in our society, and that almost all needs are becoming the targets of intensive advertising by suppliers of answers to those needs – 'needmongers' who profit from persuading us all to try to use material means to meet nonmaterial needs. That is futile and silly.

The fundamental error of economics with regard to the satisfaction of vital needs has been well described by Mary Clark (1989, p 343). She observes that economics is very capable of making apples and pears and heart pacemakers mutually comparable, but fails to make any of these goods comparable with mother's love. As we try to develop a sustainable civilisation, we have to relearn how to appreciate all the non-economic benefits and satisfactions (and to recognise that their destruction often follows being oversupplied with economic goods).

One clue to realising immaterial values – of highly positive material significance – was introduced by Ruth Benedict (1934, p 21) by the term 'synergy'. In a highly synergised society, the good of the individual coincides with the good of the whole, because the social institutions create this situation. Helping others, far from being a personal sacrifice, is a perceived personal benefit. This contrasts sharply with the anthropological outlook of modern economics, the roots of which go back to Thomas Hobbes's depiction of man as greedy, selfish and aggressive. For further study see Mary Clark's *Ariadne's Thread* (1989), notably pp 176–181.

14.2 Informal Sector Wealth and Civilisation

Jobs in the formal economy will remain highly attractive for most people able to work. However, well-being need not be eternally married to what today is called a job. The meaning of jobs will continue to change. In particular, the specific benefits that make jobs so highly attractive (regular income, social security and the feeling of being needed) may gradually be decoupled from professional employment.

Despite all the advances of the service economy, the volume of needed professional work is diminishing as a result of increased labour productivity (Rifkin, 1995). Moreover, one of the main attractions of formal jobs, the taking of social security, can gradually be decoupled from the job. Ecological tax reform can be designed in such a way that social security benefits are less dependent than they are today on regular payments by employers and employees (but are dependent on the existence of resource taxes). Also, an increasing number of people are fortunate enough to inherit sizeable amounts of money, and out-of-job earnings, solidarity among friends and family relations make people less dependent on job income. All these developments will allow people to discover that jobs are not *per se* a delight. Many people will therefore become more selective. They will accept part-time jobs. And they are likely to rediscover the pleasures of the informal sector.

How dare we speak of the pleasures of the informal sector? All of modern development was guided by a conscious desire to overcome the informal sector. It was seen as backward, the depressingly dominant sector in the poor and boring village economy of yesteryear. The economic success story of modern times was characterised by an increasing division of labour, the professionalisation of all trades (even to the point that Ivan Illich calls 'radical monopoly' in which the most basic human activities can no longer be carried on by ourselves without professional help), enormous gains in labour efficiency, growing needs for transport, the dominance of money both as an instrument of measuring exchange values and of assessing a person's success. And the triumph of the modern economy was characterised by an erosion of the informal sector, or 'subsistence economy' as it was contemptuously labelled.

In the glorification of this success story, it is sometimes forgotten that even now the formal economy would be totally helpless if the informal sector did not still exist. Sleeping, eating, loving and bringing up children are not subordinate activities we could do without, but the indispensable foundation of all human existence. Economic theory has a shocking tendency to repress this simple fact (Weizsäcker, 1994, p 197).

At any rate, a linear continuation on the road of professionalisation and monetarisation seems neither possible nor desirable. The time may have come to recognise what was lost with the erosion of the informal sector.

This is also the insight of Orio Giarini and Patrick Liedtke in their forthcoming Report to the Club of Rome on the future of work. They explicitly call for recognising the productive value of activities which contribute to the overall wealth, even when these activities do not fall or only partially fall within the statistics of the present national accounting system. The report anticipates very high variation among individuals in paid work patterns and hence a broad spectrum of options for unpaid work, which by definition is encompassed by the informal sector.

A snug harbour for children in their time of growing up and exploring the world has always been an important aspect of the informal sector. It's not only a stable family that provides such a safe haven; the neighbourhood, the church or the primary school can greatly contribute. In societies dominated by the informal sector, the haven quite naturally persists beyond adolescence and is without much ado transferred to the next generation. But it tends to be lost in a society where the informal sector is sacrificed to the money economy. The modern ills of loneliness, unrest, vandalism, drug addiction and related crime may have much to do with the decline of the informal sector.

If people are given the chance to regain that harbour while still participating in the money economy to a sufficient extent, an increasing number, we suspect, will seize the opportunity.

They may simply take greater delight in inefficiently producing strawberry yoghurt in the family or in the neighbourhood (Chapter 3, p 117), insulating their own homes (Chapter 1, p 15), doing furniture maintenance and repair (Chapter 2, p 70) or doing for themselves many other things related to the Factor Four revolution. They would certainly not see themselves as backward. And their children might grow up in a much more agreeable environment.

14.3. MARKETS ARE NOT A SUBSTITUTE FOR ETHICS, RELIGION AND CIVILISATION

Much of this book has explained how the careful application of correct economic principles can make economics once more truly what its Greek etymology suggests – 'management of the household'. This is very different from the casual misapplication of economic ideas to suppose that the economic process is simply a disembodied, endlessly circular flow of exchange value between production and consumption, yet not embedded in a flow of physical resources from depletion to pollution. This common fallacy is like trying (as the economist Herman Daly puts it) to understand an animal in terms only of its circulatory system, without noticing that it also has a digestive tract which ties it firmly to its environment at both ends. It is this fallacy that often misleads economic fundamentalists to treat soil like dirt, living things as dead, nature as a nuisance, thousands of millions of years' worth of design experience as casually discardable and the future as worthless (at a 10 per cent real discount rate).

Correct economic theories are often misused in other ways as well:

- Expressing economic output as simply the sum of the market value of goods, services, bads and nuisances sold for money, excluding everything that has no price (such as serving and caring for loved ones) and everything that is priceless. This definition is surely a bizarre way of expressing the sum of human happiness or even of material satisfaction. By most measures (Cobb and Cobb, 1994), the latter has been falling since the early 1970s (Chapter 12), since growth in GDP has been more than swallowed up by the 'remedial costs' of coping with depletion, pollution, deferred maintenance or renewal, and social disorder.
- Assuming that depletion and pollution cost nothing (depletion is usually counted only as extraction cost and much pollution as zero), however dearly they may cost our descendants – a common and widely sanctioned theft from the future. (As editor Hugh Nash put it, how can we better pay tribute to our children's boundless technological ingenuity than to make sure they'll need it?)
- Treating consumption of capital as income – despite the economist Hicks's correct definition of income as the maximum

amount you can consume over a given period without being worse off than when you started. (We wish we could figure out how to 'develop' our bank account by withdrawing money from it; yet this is exactly what most economic 'development' does. Cutting down trees, destroying fisheries or mining ores adds income, but debits no corresponding withdrawals to the global balance sheet, because there isn't one.)

- Including only what's countable, not what really counts.
- Assuming infinite substitutability of artificial capital for natural capital. (On this dubious theory, some economists urge us to deplete any available resource at the economically optimal rate, typically equal to the real discount rate – approximately how much interest, beyond inflation, money can earn if wisely invested – because after the resource, be it whales or rainforests, is all gone, the money can be reinvested in some other resource that will fully substitute for the one just depleted. If this were true, economists would be God. Economics is certainly the state religion, and its high priests and acolytes are well entrenched in their rituals, but some of us have other ideas about true gods and false idols.)
- Improperly treating buying and selling bids as equivalent. (A government agency that wants to build a motorway through your back garden should first ask how much compensation you'll require, and if you reply that your garden isn't for sale at any price, that should end the matter. Unfortunately, dishonest agencies often reverse the question into 'How much would you pay us in order not to lose your garden?' The answer to that question of course depends on your wealth, not just your willingness to pay; and there are an infinite number of assaults you can be asked to forestall in this manner, quickly exhausting even near-infinite wealth. The reversal of the question is both economically improper and immoral, but is commonly done, because otherwise – given that different people have different values – there is no theoretical basis for using cost-benefit analysis for public policy decisions such as whether and where to build motorways.)
- Discounting things that can't be banked at interest, such as life, liberty and the pursuit of happiness. (British officials used to use the Treasury's high discount rates – saying that after factoring out inflation, each £1 of benefit next year is worth spending only 90 pence today to achieve – to justify jerry-building blocks of flats that could fall on your head in 20 years, because a 20-year risk is hardly worth anything: specifi-

cally, 15 per cent of its value today. They stopped, it is said, when their Dutch colleagues pointed out that if Holland had used that shortsighted philosophy, it would have been under water long ago.)

In principle, all these abuses of sound economic theory can be corrected. But using markets for what they do extremely well – to allocate scarce resources optimally over the short term – cannot enable markets to do things that they cannot do at all. For example, markets

- cannot tell you what the right size is (as the economist Herman Daly reminds us, 'A boat that tries to carry too much weight will still sink even if that weight is optimally allocated');
- do not reveal when legitimate needs are being subordinated and sacrificed to extravagant wants;
- don't say when to stop or how much is enough (Ecclesiastes warned us that 'He that hath silver shall not be satisfied with silver, nor he that hath abundance with increase; this is also vanity').

Economics does not fully reflect human purpose, because human purpose far surpasses greed and envy. But in the quest for an ever grosser National Product, it is all too easy to forget the ends that economic tools were constructed to serve. Our public discourse about economic goals will continue to substitute empty slogans for real values until we remember, in Dana Meadows's words, that:

> *A sustainable society would be interested in qualitative development, not physical expansion. It would use material growth as a considered tool, not a perpetual mandate. It would be neither for nor against growth. Rather, it would begin to discriminate kinds of growth and purposes for growth. Before this society would decide on any specific growth proposal, it would ask what the growth is for, and who would benefit, and what it would cost, and how long it would last, and whether it could be accommodated by the sources and sinks of the planet.*

All societies today, we believe, hunger for leaders who ask these questions – leaders who extend, as Robert Gilman urges, the Golden Rule through time, doing unto future generations of living beings as we would have done unto us. To heal our societies, create justice and restore the earth, we will need to rethink fundamentally our relationship to Creation, to our own capacities and purposes, and to our children.

Public life offers myriad recent examples of basic confusions between the creation of wealth and the pursuit of happiness; between profit opportunities and basic rights; between private gains and public goals. After all, economic efficiency is only a means, not an end. Markets are meant to be efficient, not sufficient; greedy, not fair. Markets were never meant to achieve community or integrity, beauty or justice, sustainability or sacredness – and they don't. If markets do something good for whales or wildness, for God or Gaia or grandchildren, that is purely coincidental. Markets, if allowed to work properly, are very good at achieving their stated goals, but those goals are far from the whole purpose of a human being. It is to seek that fuller purpose that we have politics, ethics and religion; and if we ever suppose these greatest achievements of the human spirit can be replaced by economics, we stand in peril of our souls.

Mullah Nasruddin was once asked which is more valuable, the moon or the sun. 'Why, the moon!', he replied. 'And why is that, great Mullah?' 'Oh, because it shines at night, when we need the light more.' Today, too, we need the light more. Productive use of the gifts we borrow from the earth and from each other can gain us more time to search. But like any tool, it can only help us a little towards, and can never substitute for, the renewal of our polity, our ethical principles and our spirituality. The resources that we need most urgently to rediscover and to use more fully and wisely are not in the physical world, but remain hidden within each one of us.

References

Afheldt, H (1994) *Wohlstand für niemand?* Munich, Kunstmann

Anderson, M S (1994) The Green Tax Reform in Denmark *J of Envir. Liability*, 2, pp 47–50

Arrhenius, Svante (1896) On the Influence of Carbonic Acid in the Air upon the Temperature on the Ground *Phil Mag* 5, 41 (251) p 237

Ayres, R U and L W Ayres (1996) *Industrial Ecology: Towards Closing the Materials Cycle* Cheltenham

Ballard, C L and S G Medema (1992) *The Marginal Efficiency Effects of Taxes and Subsidies in the Presence of Externalities: A Computational General Equilibrium Approach* East Lansing MI, Michigan State University

Bangladesh Centre for Advanced Studies (1994) *Vulnerability of Bangladesh to Climate Change and Sea Level Rise* Summary Report Dhaka, Bangladesh, BCAS

Barbir, F, T N Veziroglu and H J Please (1990) Environmental Damage Due to Fossil Fuel Use *Int J Hydrogen Energy*, 15, pp 739–749

Barney, G 1980 *The Global 2000 Report to the President* Harmondsworth, Penguin

Barnola, J M, D Raynaud, Y S Krokotkevitch and C Lorius (1987) Vostok Ice Core: A 160,000-Year Record of Atmospheric CO_2 *Nature* 329, pp 408–414

Benedick, R E (1991) *Ozone Diplomacy: New Directions in Safeguarding the Planet* Cambridge MA, Harvard University Press

Benedict, R (1934) *Patterns of Culture* Boston, Houghton Mifflin

Bleischwitz, R and H Schütz (1992) *Unser trügerischer Wohlstand: Ein Beitrag zur deutschen Ökobilanz* Wuppertal Institute

Böge, S (1993) Erfassung and Bewertung von Transportvorgängen: Die produktbezogene Transportkettenanalyse, in D Läpple (Ed) *Güterverkehr, Logistik und Umwelt* Berlin, Edition Sigma

Brown, L, D Dennison, C Flavin et al (1995) *State of the World 1995* London, Earthscan

— (1992) Environment and Trade as Partners in Sustainable Development *The American Journal of International Law* 86, p 728 ff

Browning, W D and D L Barnett (1995) *A Primer on Sustainable Building* Snowmass, CO, Rocky Mountain Institute

Brundtland Report: see WCED

Buitenkamp, M, H Venner and T Wams (Eds) (1992) *Action Plan Sustainable Netherlands* Amsterdam, Milieudefensie

BUND, Misereor and Wuppertal Institut (1996) *Zukunftsfähiges Deutschland* Basel, Birkhäuser

Burwitz, H, H Koch and T Krämer-Badoni (1992) *Leben ohne Auto: Neue Perspektiven für eine menschliche Stadt* Reinbek, rororo

Cairncross, F (1991) *Costing the Earth* Boston, Harvard Business School Press and London, Economist Books

Carson, R (1962) *Silent Spring,* Greenwich, CT

Clark, D (1995) AutoDesk to Publish Digital Drawings of Manufacturers' Parts for Engineers, *Wall Street Journal*, 3 March

Clark, M (1989) *Ariadne's Thread: The Search for New Modes of Thinking* London, Macmillan

Club of Rome (1991) *see* King, A and B Schneider

— (1993) *Growth, Competitiveness and Employment* White Paper, Luxembourg, EC Publications

Cobb, C W and J Cobb Jr (1994) *The Green National Product* Lanham MD, University Press of America

Commission of the European Communities (1992) *Towards Sustainability: A European Community Programme of Policy and Action in Relation to the Environment and Sustainable Development* Brussels, EC Commission

Commission of the European Communities (1993) *Growth, Competitiveness and Employment* White Paper, Luxembourg, EC Publications

Corbett, M N (1981) *A Better Place to Live: New Designs for Tomorrow's Communities* Emmaus, PA, Rodale Press

Corden, W M (1974) *Trade Policy and Economic Welfare* Oxford, OUP

Costanza, R (1992) *Ecological Economics: The Science and Management of Sustainability* New York, Columbia University Press

Daly, H (1991) *Steady State Economics* 2nd ed, Washington, Island Press

Daly, H and J B Cobb (1989) *For the Common Good: Redirecting the Economy Toward Community, the Environment and a Sustainable Future* Boston, Beacon Press

D'Aveni, R (1994) *Hypercompetition* New York, Free Press

Dawkins, R (1976) *The Selfish Gene* Oxford, OUP

De Andraca, R and K F McCready (1994) *Internalizing Environmental Costs to Promote Eco-Efficiency* Stockholm, Tomorrow Publications

de Moor, A (1997) *Subsidising Unsustainable Development: Undermining the Earth with Public Funds* Toronto, Earth Council

Dearien, J A and M M Plum (1993) 'The Capital, Energy and Time Economics of an Automated, On-Demand Transportation System' Twenty-eighth Intersociety Energy Conversion Engineering Conference, 8–13 August 1993, Atlanta, Georgia (copies may be obtained from J A Dearien, c/o Lockheed INEL, PO Box 1625, Idaho Falls, ID 83415-3765, USA)

Diederichs, C J and F J Follmann (1995) Ressourcenschonendes Bauen eröffnet Perspektiven für den Umweltschutz der Zukunft *Bauwirtschaft* 49, pp 32–39

Duchin, F and G M Lange (1994) *The Future of the Environment* Oxford and New York, OUP

Ederer, G and P Ederer (1995) *Das Erbe der Egoisten* Munich, Bertelsmann

Eichner A S (1995) The Lack of Progress in Economics *Nature* 313, pp 427–428

Ekins, P, C Folke and R Costanza (1994) Trade, Environment and Development: the Issues in Perspective *Ecological Economics* 9, pp 1–12

Ekins, P and M Max-Neef (eds) (1992) *Real-Life Economics: Understanding Wealth Creation* London, Routledge

Enquête-Kommission 'Schutz des Menschen und der Umwelt' (1994) *Die Industriegesellschaft gestalten: Perspektiven für einen nachhaltigen Umgang mit Stoff- und Materialströmen* Bonn

Feist, W and J Klien (1994) *Das Niedrigenergiehaus* Heidelberg

Fickett, A P, C W Gellings and A B Lovins (1990) Efficient Use of Electricity *Scientific American* September 1990, pp 65–74

Fowler, C and P Mooney (1990) *Shattering: Food, Politics and the Loss of Genetic Diversity* Tucson, University of Arizona Press

Frank, J E (1989) *The Costs of Alternative Development Patterns: A Review of the Literature* Washington

Fussler, C and P James(1996) *Driving Eco Innovation* London, Pitman

Gabor, D, U Colombo *et al* (1976) *Beyond the Age of Waste* A Report to the Club of Rome. Oxford, Pergamon

GATT (1992) *Trade and the Environment* Geneva, GATT/WTO Sectretariat

Giarini, O and W R Stahel (1993) *The Limits to Certainty: Facing Risks in the New Service Economy* Dordrecht, Kluwer Academic

Gillies, A M (1994) *The Greening of Government Taxes and Subsidies: Protecting the Environment and Reducing Canada's Deficit* Winnipeg, International Institute for Sustainable Development

Goldsmith, E and N Hildyard (1985–1989) *The Social and Environmental Effects of Large Dams* 3 Vols, Camelford, Cornwall, Ecosystems

Goldsmith, J (1995) *The Trap* London, Macmillan

Gore, A (1992) *Earth in the Balance: Ecology and Human Spirit* Boston and London, Houghton Mifflin

Greenpeace (ed) (1995) *Der Preis der Energie-Plädoyer für eine Ökologische Steurreform* Munich, Beck

Groupe de Lisbonne (1995) *Limites de la compétitivité* Paris, La Découverte

Hawken, P (1994) *The Ecology of Commerce: A Declaration of Sustainability* New York, Harper Business

Henderson, H (1991) *Paradigms in Progress*, Indianapolis, Knowledge Systems

Henderson, H (1996) *Building A Win-Win World* San Francisco, Berret-Koehler

Hennicke, P and D Seifried (1996) *Das Einsparkraftwerk* Basel, Birkhäuser

Houghton D and D Hibberd (1993) Packaged Rooftop Air Conditioners Tech Update TU-93-1, E source (Boulder CO 80302), January

Howe, W, M Shepard, A B Lovins *et al.*, *Drivepower Technology Atlas*, E source, August 1993

Illich, I (1972) *Tools for Conviviality* New York, Harper and Row

IPCC (Intergovernmental Panel on Climate Change) (1990) *The IPCC Scientific Assessment*, J T Houghton, G T Jenkins and J J Ephraums, (Eds), Cambridge, CUP; new edition 1996

IPSEP, see Krause, Florentin *et al.*

IUCN (International Union for the Conservation of Nature; since 1990: World Conservation Union) (1981) *Caring for the Earth: A Strategy for Sustainable Living* London, Earthscan (this book also summarises the 'World Conservation Strategy' published by the same organisations at Geneva in 1980)

Jackson, T and N Marks (1994) *Measuring Sustainable Economic Welfare – A Pilot Index 1950–1990* Stockholm, Stockhom Environment Institute

Jackson, W (1980) *New Roots of Agriculture* San Francisco, FoE; see also the Land Institute's Annual Reports Salina, K67401, USA

Jeavons, J (1971) *How to Grow More Vegetables Than You Ever Thought Possible on Less Land Than You Can Imagine* Willits CA

Jeavons, J and C Cox (1993) *Lazy-Bed Gardening* Willits CA

Jencks, C (1990) Post-Modernism: Between Kitsch and Culture, In: *Architectural Design, Post-Modernism on Trial*, pp 30–31, London, Academy Editions

Jouzel, J *et al.* (1987) Vostok Ice Core: A Continuous Isotope Temperature Record over the Last Climatic Cycle (160,000 years) *Nature* 329, pp 403–408

Kempton, W, D Feuermann *et al.* (1992) 'I Always Turn It on Super': User Decisions about When and How to Operate Room Air Conditioners *Energy and Buildings* 18 (3) pp 177–191

King, A and B Schneider (1991) *The First Global Revolution: From the Problematique to the Resolutique*, by the Council of the Club of Rome, New York, Pantheon

Klingholz, R (1994) *Wahnsinn Wachstum: Wieviel Mensch erträgt die Erde?* Hamburg, Geo Buch

Klüting, R (1995) Der Faktor 10 Club, *Bild der Wissenschaft* March, pp 78–80

Kohn, A (1986) *No Contest: The case against competition* Boston, Houghton Mifflin

— (1990) *The Brighter Side of Human Nature: Altruism and Empathy in Everyday Life* New York, Basic Books

Korten, D (1995) *When Corporations Rule the World* San Francisco, Berrett-Koehler

Knoflacher, H (1995) Economy of Scale: Die Transportkosten and das Ökosystem *Gaia* 4, pp 100–108

Kranendonk, S and S Bringezu (1993) First Estimates of the Material Intensity of Orange Juice *Fresenius Environmental Bulletin* 2, pp 455–460

Krause, F, J Hooney *et al.* (1993–1996) *Energy in the Greenhouse*; Vol 2, Report of the International Project for Sustainable Energy Paths (IPSEP) to the Dutch Ministry of the Environment, El Cerrito, CA, IPSEP

Liedtke, C (1993) Material Intensity of Paper and Board Production in Western Europe *Fresenius Environmental Bulletin* 2, pp 461–466

Liedtke, C and T Merten (1994) *MIPS, Resource Management and Sustainable Development*, pp 163–173, Proceedings of the Second International Conference on 'The Recycling of Metals' Amsterdam

Lorius, C *et al.* (1985) A 150,000-Year Climatic Record from Antarctic Ice *Nature* 316, pp 591–596

Lovins, A B (1988) Negawatts for Arkansas, 3 vols, RMI Publ no U88-30

Lovins, A B (1990a) How a Compact Fluorescent Lamp Saves a Ton of CO_2, RMI Publ no 90-5

— (1992) Energy-Efficient Buildings: Institutional Barriers and Opportunities, E source (Boulder CO), Strategic Issues Paper no 2, November

— (1993) What an Energy-Efficient Computer Can Do, Rocky Mountain Institute Publ. no E93-20

— (1995) The Super-Efficient Passive Building Frontier *ASHRAE J* June (1995); also RMI Publ no E 95-28

— (1996) Negawatts: Twelve Transitions, Eight Improvements and One Distraction *Energy Policy* (UK), April; also RMI Publ no U96-11

Lovins, A B, M M Brylawski, D R Cramer, T C Moore (1996) *Hypercars: Materials, Manufacturing and Policy Implications*, The Hypercar Centre, Rocky Mountain Institute, Snowmass CO 81654-9199

Lovins, A B and L H Lovins (1995) Reinventing the Wheels *Atlantic Monthly* 271, pp 75–81

Lünzer, I (1979) Energiefragen im Umwelt and Landbau. Reprinted (1991) in: H Vogtmann (Ed) *Ökologische Landwirtschaft*, pp 277–302, Karlsruhe, C F Müller

— (1992) Energiebilanzen in der Landwirschaft bei unterschiedlichen Wirtschaftweisse. In Stiftung Ökologie und Landbau (Ed.) Sinnvoller Umgang mit Energie auf dem Bauernhof. Bad Dürkheim, SÖL, pp 15–34

MacKenzie, J J, R Dower and D Chen (1992) *The Going Rate: What It Really Costs to Drive* Washington, World Resources Institute

Maddox, J (1972) *The Doomsday Syndrome* London, Macmillan

Mander, Jerry and E Goldsmith (1995) *The Case against the Global Economy* San Francisco, Sierra Club Books

Masuhr, K P, S Schärer and H Wolff (1995) *Energieverbrauch: Kostenwahrheit ohne staat? Die externen Kosten der Energieversorgung und Internalisierung.* Stuttgart, Metzlen Poeschel

Meadows, D, D Meadows and J Randers (1992) *Beyond the Limits: Confronting Global Collapse, Envisioning a Sustainable Future* Post Mills VT, Chelsea Green

Meadows, D H, D L Meadows, J Randers, C W Behrens (1972) *The Limits To Growth* New York, Universe Books

Meyer, N I, O Benestad, L Emborg *et al.* (1993) Sustainable Energy Scenarios for the Scandinavian Countries *Renewable Energy* 3, pp 127–136

Meyer, N, and J S Nørgård. Planning Implications of Electricity Conservation: The Case of Denmark, in: T B Johansson *et al.* (Eds) *Electricity: Efficient End-Use and New Generation Technologies, and Their Planning Implications* Lund, Lund University Press

Mills, E (1989) *An End-Use Perspective on Electricity Price Responsiveness*, Lund, Lund University Press

Nelson, K E (1993) Dow's Energy/WRAP Contest, Lecture given at the 1993 Industrial Energy Technology Conference, Houston, 24–25 March. (Copies available from E source, Boulder CO 80302-5114 or Kentech, Baton Rouge, LA, USA

Neumann-Mahlkau, P (1993) Acidification by Pyrite Weathering on Mine Waste Stockpiles, Ruhr District, Germany *Engineering Geology* 34, pp 125–134

Nordhaus, W (1990) Count Before You Leap: Economics of Climate Change *Economist*, July 1990

— (1993) *Managing the Global Commons* Cambridge MA, MIT Press

Nørgård, J S (1989) Low Electricity Appliances – Options for the Future, in: T B Johansson *et al.* (Eds) *Electricity: Efficient End-Use and New Generation Technologies, and Their Planning Implications* Lund, Lund University Press

Olivier, D (1992) *Energy Efficiency and Renewables: Recent Experience on Mainland Europe* Herefordshire, Energy Advisory Associates

Panayotou, T (1993) *Green Markets: The Economics of Sustainable Development* San Francisco, ICS Press

Petersen, M (1994) *Ökonomische Analyse des Car-sharing* Berlin, Stattauto

Petersen, R and K-O Schallaböck (1995) *Mobilität für morgen: Chancen einer zukunftsfähigen Verkehrspolitik* Basel, Birkhäuser

Pigou, A C (1920) *The Economics of Welfare* London, Macmillan

PCSD (President's Council for Sustainable Development) (1996) Sustainable America: A new consensus for prosperity, opportunity and a healthy environment for the future. Washington DC

Prince of Wales, HRH The (1989) *A Vision of Britain* New York, Doubleday

Rabinovich, J and J Leitman (1996) Urban Planning in Curitiba *Scientific American* March, pp 46–53

Read, P (1994) *Responding to Global Warming: The Technology, Economics and Politics of Sustainable Energy* London, Zed Books

Rechsteiner, R (1993) Sind hohe Energiepreise volkswirtschaftlich ungesund? *Gaia* 2, pp 310–327

RERC (Real Estate Research Corporation) (1974) The Costs of Sprawl: Environmental and Economic Costs of Alternative Residential Development Patterns at the Urban Fringe, 3 vols, Report for the President's Council on Environmental Quality, Washington

Rees, W and M Wackernagel (1994) Ecological Footprints and Appropriated Carrying Capacity: Measuring the Natural Capital Requirements of the Human Economy, in: A M Jannson *et al.* (Eds) *Investing in Natural Capital: The Ecological Economics Approach to Sustainability* Washington, Island Press

Renner, M (1992) *Budgeting for Disarmament: The Costs of War and Peace* (Worldwatch Paper No 122) Washington DC, Worldwatch Institute

Repetto, R, R Dower et al (1992) *Green Fees: How a Tax Shift Can Work for the Environment and the Economy* Washington, DC, World Resources Institute

Rifkin, J (1995) *The End of Work: The Decline of the Global Labor Force and the Dawn of the Post-Market Era* New York, Putnam

Ripl, W (1994) *Management of Water Cycle and Energy Flow for Ecosystem Control: The Energy-Transport-Reaction (ETR) Model* Amsterdam

RMI (Rocky Mountain Institute) (1994) *Water Efficiency: A Resource for Utility Managers, Community Planners and Other Decisionmakers*, (in cooperation with US EPA) 4th ed, November, Snowmass, CO

— (1995) *A Primer on Sustainable Building* RMI Publ no D95-2

Romm, J J (1994) *Lean and Clean Management: How to Boost Profits and Productivity by Reducing Pollution* New York, Kodansha International

Romm, J J and W D Browning (1994) *Greening the Building and the Bottom Line: Increasing Productivity through Energy-Efficient Design* Snowmass, CO, RMI Publ D4-27

Scherhorn, G (1994) Die Untersättlichkeit der Bedürfnisse and der kalte Stern der Knappheit, in: B Biervert and M Held (Eds) *Das Naturverständnis der Ökonomik* Frankfurt; see also: G Scherhorn (1994) Egoismus oder Autonomie: Über die Beschränktheit des Eigennutzprinzips, in: T L Heck (Ed) *Das Prinzip Egoismus* Tübingen, Nous Verlag

Schmidheiny, S with the Business Council for Sustainable Development (1992) *Changing Course: A Global Perspective on Development and the Environment* Cambridge MA, MIT Press

Schmidt-Bleek, F (1994) Carnoules Declaration of the Factor Ten Club, Wuppertal Institute

Shiva, V (1992) *Staying Alive Women, Ecology and Development* London, Zed Books

Stahel, W and E Gomringen (1993) (International Design Forum/IFG) *Gemeinsam nutzen statt einzeln verbrauchen* Giessen, Anabas Verlag

Third World Network (1995) *The Need for Greater Regulation and Control of Genetic Engineering: A Statement by Scientists Concerned about Current Trends in the New Biotechnology* Penang, Malaysia, Third World Network

Tobin, J (1994) A Tax on International Currency Transactions, in: UNDP (1994) *Human Development Report*, p 70

Tooley, M J (1989) Global Sea Levels: Floodwaters Mark Sudden Rise *Nature* 342, pp 20–21

Tsuchiya, H (1994) Energy Analysis of Daily Activities *Energy for Sustainable Development* 1, pp 39–43

UNDP (1990) *Human Development Report* New York, UNDP

UNDP (1994) *Human Development Report* New York, UNDP

Vale, B and R Vale (1991) *Green Architecture: Design for and Energy-Conscious Future* Bullfinch Press, Little Brown, Boston, pp 156–168

van Dieren, W (Ed) (1995) *Taking Nature Into Account: A Report to the Club of Rome* New York, Springer Verlag

Vester, F (1995) *Crashtest Mobilität* Munich, Heyne

WCED (World Commission on Environment and Development) (1987) *Our Common Future* (The Brundtland Report) Oxford, OUP

Weiss, E B (1989) *In Fairness to Future Generations: International Law, Common Patrimony and Intergenerational Equity* Tokyo, UNU and Dobbs Ferry

Westin, R A (1997) *Environmental Tax Initiatives and Multilateral Trade Agreements* Dordrecht/Cambridge, Mass, Kluwer Law International

Weizsäcker, C von and E U von Weizsäcker (1978) Recht auf Eigenarbeit statt Pflicht zum Wachstum *Scheidewege* 9, pp 221–234

Weizsäcker, E U von and J Jesinghaus (1992) *Ecological Tax Reform* London, Zed Books

Weizsäcker, E U von (1994) *Earth Politics* London, Zed Books

Weterings, R and J B Opschoor (1992) *The Ecocapacity as a Challenge to Technological Development* Rijswijk

Wilson, E (1992) *The Diversity of Life* Cambridge MA, Belknap Press

World Business Council for Sustainable Development (WBCSD) (1997) *Environmental Performance and Shareholder Value* J Blumberg, Å Korsvold and G Blum (eds); Poole, Dorset; E & Y Direct

World Energy Council (1993) *Energy for Tomorrow's World: The Realities, the Real Options and the Agenda for Achievement* London, Kogan Page

Worldwatch Institute see Brown, Lester (Ed)

Wuppertal Institute (1995) *Towards Sustainable Europe* Luton, Friends of the Earth

Yergin, D (1991) *The Prize: The Epic Quest for Oil, Money and Power* London, Simon and Schuster

Young, J E (1992) *Mining the Earth* Worldwatch Paper no 109 Washington, Worldwatch Institute

Index

Page numbers in *italics* refer to figures